THE FRONTIERS COLLECTION

THE FRONTIERS COLLECTION

The books in this collection are devoted to challenging and open problems at the forefront of modern science, including related philosophical debates. In contrast to typical research monographs, however, they strive to present their topics in a manner accessible also to scientifically literate non-specialists wishing to gain insight into the deeper implications and fascinating questions involved. Taken as a whole, the series reflects the need for a fundamental and interdisciplinary approach to modern science. Furthermore, it is intended to encourage active scientists in all areas to ponder over important and perhaps controversial issues beyond their own speciality. Extending from quantum physics and relativity to entropy, consciousness and complex systems – the Frontiers Collection will inspire readers to push back the frontiers of their own knowledge.

Other Recent Titles

Weak Links
Stabilizers of Complex Systems from Proteins to Social Networks
By P. Csermely

The Biological Evolution of Religious Mind and Behaviour
Edited by E. Voland and W. Schiefenhövel
Particle Metaphysics

Principles of Evolution
by Meyer-Ortmanns and Thurner

The Physical Basis of the Direction of Time
By H.D. Zeh

Mindful Universe
Quantum Mechanics and the Participating Observer
By H. Stapp

Decoherence and the Quantum-To-Classical Transition
By M. Schlosshauer

The Nonlinear Universe
Chaos, Emergence, Life
By A. Scott

Symmetry Rules
How Science and Nature are Founded on Symmetry
By J. Rosen

Quantum Superposition
Counterintuitive Consequences of Coherence, Entanglement, and Interference
By M.P. Silverman

Martin Brinkworth • Friedel Weinert
Editors

Evolution 2.0

Implications of Darwinism in Philosophy and the Social and Natural Sciences

Editors
Dr. Martin Brinkworth
University of Bradford
School of Medical Sciences
Great Horton Road
BD7 1 DP Bradford
United Kingdom
M.H.Brinkworth@bradford.ac.uk

Dr. Friedel Weinert
University of Bradford
Division of Humanities
Great Horton Road
BD7 1DP Bradford
United Kingdom
f.weinert@bradford.ac.uk

ISSN 1612-3018
ISBN 978-3-642-20495-1 e-ISBN 978-3-642-20496-8
DOI 10.1007/978-3-642-20496-8
Springer Heidelberg Dordrecht London New York

Library of Congress Control Number: 2011939306

Printed on acid-free paper

Springer is part of Springer Science+Business Media (www.springer.com)

Foreword: The Debate over Darwinism

The year 2009 did not lack for Darwin anniversary meetings, all over the world. Yet the conference that took place in the northern city of Bradford – where most of the papers collected in this splendid volume were originally presented – marked an especially fitting tribute. For Bradford is really where the story started. Not, of course, the story of how Darwin came to develop his evolutionary ideas, or to compose *On the Origin of Species by Means of Natural Selection* in a way that eventually made those ideas persuasive to the scientific community. What began in the Bradford region is the tradition of creative disagreement about what those ideas mean.

It is too little remembered, even locally, that Darwin was in the village of Ilkley – just 20 minutes north of the city – when the *Origin* was published on 24 November 1859. He first laid eyes on the *Origin* in Ilkley in early November. 'I am *infinitely* pleased and proud at the appearance of my child', Darwin wrote back to his London-based publisher, John Murray. And it was during Darwin's visit that, with the help of the village post office, he launched himself on the hard work of converting or, as one of his friends joked, perverting his peers to the new ideas. Mike Dixon and I have told the full story of Darwin's 9-week stay in our book *Darwin in Ilkley* [1]. Here I want only to sketch the background to Darwin's trip up north at such a consequential moment, and also to examine briefly a part of that initial debate – notably to do with the evolution of mind, the question of purpose or teleology in evolution, and the vexed matter of evolution's political or ideological implications.

What brought Darwin to Ilkley in the autumn of 1859? The answer is straightforward: he came for the 'water cure', or 'hydropathy' as it was more fancily known. This was a fashionable alternative therapy of the day. Devotees of the cure subjected themselves to a regime of cold baths, wet sheets and copious drinking of cold water, combined with simple eating and outdoor walks. By mid-1859 Darwin, then 50 years old and a man who had spent much of his adulthood suffering from a mysterious ailment, had become a fan of the cure, and the visit to Ilkley was his treat to himself for having slaved over the proofs of the *Origin* the previous months. He arrived on 4 October 1859 with his health broken, and left on 7 December feeling, for him, not too bad. '[D]uring great part of day I am wandering on the hills, and trying to inhale health,' he wrote to the cleric, naturalist and man of letters Charles Kingsley on 30 November. And in mid-December, now back in Kent, he wrote to his brother Erasmus: 'The latter part of my stay at Ilkley did me much good.'

For a couple of weeks at either end of the visit Darwin stayed by himself at Ilkley's grand hydropathic hotel (now luxury flats). In between he was joined by his wife and children, residing with them just down the road from the hotel, in another building which still stands. His family left on 24 November – publication day for the *Origin*. Although Darwin did manage to get some rest, the book was a constant presence throughout the 9 weeks. In Ilkley he made final-final changes to the text, decided on the people who were to receive complimentary advance copies, and even, after publication, made the small but significant changes that went public in January 1860 within the second edition of the *Origin*. It was also here that he awaited the judgement of the scientific world on the book. There were the newspaper and journal reviewers, of course, and the recipients of the advance copies. But no judgement meant more to him that that of his friend and mentor Sir Charles Lyell, who had spent the summer of 1859 reading copies of the corrected proofs.

Darwin's Ilkley correspondence with Lyell, which started almost immediately after Darwin's arrival, preserves a debate that easily ranks as the deepest and most important that Darwin ever engaged in over his book. Lyell was one of the greatest nineteenth-century British geologists, who taught that earthly change has always been a matter of the slow, gradual accumulation of the effects of the small-scale causes of change observed today: wind, rain, earthquakes and so on. Darwin was a Lyellian from the time he was a young man on the *Beagle* voyage; soon after the voyage, the discipleship became a friendship. For Darwin, Lyell towered over other naturalists – he was Darwin's 'Lord Chancellor', as Darwin once put it in a letter – and so Lyell's response meant a great deal, personally but also strategically, in that, Darwin reckoned, where Lyell led, others would follow.

The letters that flowed between Lyell and Darwin throughout October and November 1859 record a searching, wide-ranging, no-holds-barred discussion of Darwin's proposals in the *Origin*. In the way that good mentors are, Lyell was encouraging and helpful in all kinds of ways. But he was no evolutionist, and so pressed Darwin very hard indeed on his arguments for an evolutionary theory that, in its emphasis on the gradually accumulating effects of processes observable today, was alarmingly Lyellian. Not least troubling about the theory for Lyell, a devout Christian, was whether the theory assigns God an implausibly small role in the species-making process. Famously, or notoriously, natural selection makes God a hypothesis of which we have no need – except, maybe, as the being who created the laws of nature behind natural selection. For Lyell, by contrast, plant and animal species were God's handiwork, down to the finest detail. As he had written near the close of his *Principles of Geology* (1830–1833): '[I]n whatever direction we pursue our researches, whether in time or space, we discover everywhere clear proofs of his Creative Intelligence, and of His foresight, wisdom, and power'.

In the Ilkley correspondence between Darwin and Lyell, one issue that brought these concerns about divine knowledge and foresight into the open was the question of evolutionary progress. Over and over again in his letters to Darwin, Lyell asked, in different ways, whether evolution by natural selection can by itself satisfactorily account for how a planet that, at one time in the past, was populated by animals no more intelligent than Lepidosiren (primitive South American fish), eventually came to support animals as intelligent as Lyells. As Lyell appreciated, natural selection is a theory of what happens when ordinary processes of reproduction meet the ordinary struggle for life. But, asked Lyell, is not the shift from something as simple as a fish to something as complex as a human extraordinary – so much so that its explanation must involve something more than ordinary processes? Perhaps, Lyell, went on, we need to make appeal to some further, extraordinary principle – a complexifying principle, a principle of progress – programmed into life from the beginning. On such a view, evolution becomes not the chancy, undirected business it was for Darwin, but the gradual unfolding or realization of God's plan, with the emergence of Man at the end as the goal, the *telos*.

Needless to say, such a view is anathema to Darwinians. They will be glad to learn that, in reply to Lyell, Darwin did not let them down. To accept the theory of natural selection as explaining the fish-to-man shift, said Darwin, all one needs to accept is that (1) some individuals are more intelligent than others, (2) at least some of that variability in intelligence is inherited, and (3) being more intelligent is an advantage in the struggle for life. Provided these conditions are met – and, Darwin thought, they obviously are – then natural selection can accumulate intelligence, with no limits. As Darwin summarized to Lyell, there is 'no difficulty in the most intellectual individuals of a species being continually selected; & the intellect of the new species thus improved...'

So: no spooky surplus principles needed. But Darwinian readers should not cheer too loudly for their hero quite yet, for Darwin went on, by way of offering Lyell persuasive evidence of selection's power to increase intelligence, to suggest that the process can be observed now 'with the races of man; the less intellectual races being exterminated...' Such passages in Darwin's writings, published and private, make for uncomfortable reading in the twenty-first century, and it is tempting to overlook them. But anniversaries should be occasions for reflecting both on what we now approve of in Darwin and what we find incorrect or even repellent.

Let us continue, however, with the fish-to-man letter; for Darwin goes on to give Darwinians something to cheer about – a statement as strong as the most ardent ones could wish for affirming Darwin's opposition to spookiness in science. He wrote to Lyell: 'I would give absolutely nothing for theory of nat. selection, if it requires miraculous additions at any one stage of descent ... I think you will be driven to reject all or admit all.' It is worth thinking about that last line, on reading through the chapters that follow, about everything from the possibility of Darwinising Lamarckian change, the prospects for Darwinian medicine, the problem of the ethical treatment of our fellow animals. To Darwin, acceptance of his theory was all or nothing; one was either with him all the way, or against him all the way. Yet in the end, he got Lyell to come with him only most of the way (Lyell never fully admitted humankind into the ordinary-evolutionary picture). For us, more than a century and a half later, and whether we are religiously inclined or not, it remains an open question whether we wish to go all the way with Darwin – and ,if we do wish to go all the way, a no less open question as to where that commitment will take us.

Centre for History and Philosophy of Science
University of Leeds
Leeds, UK

Gregory Radick

Reference

1. M. Dixon, G. Radick: Darwin in Ilkley. The History Press, Stroud (2009)

Contents

Introduction 1
Martin Brinkworth and Friedel Weinert

Part I Darwinism in Approaches to the Mind

The Embodiment of Mind 11
Gerald M. Edelman

Depression: An Evolutionary Adaptation Organised Around the Third Ventricle 23
Colin A. Hendrie and Alasdair R. Pickles

Does Depression Require an Evolutionary Explanation? 33
Sarah Ashelford

A Darwinian Account of Self and Free Will 43
Gonzalo Munevar

The Problem of 'Darwinizing' Culture (or Memes as the New Phlogiston) 65
Timothy Taylor

Part II Impact of Darwinism in the Social Sciences and Philosophy

Evolutionary Epistemology: Its Aspirations and Limits 85
Anthony O'Hear

***Angraecum sesquipedale*: Darwin's Great 'Gamble'** 93
Steven Bond

Darwinian Inferences 111
Robert Nola and Friedel Weinert

Breaking the Bonds of Biology – Natural Selection in Nelson and Winter's Evolutionary Economics 129
Eugene Earnshaw-Whyte

The Ethical Treatment of Animals: The Moral Significance of Darwin's Theory 147
Rob Lawlor

Part III Philosophical Aspects of Darwinism in the Life Sciences

Is Human Evolution Over? 167
Steve Jones

Evolutionary Medicine 177
Michael Ruse

The Struggle for Life and the Conditions of Existence: Two Interpretations of Darwinian Evolution 191
D.M. Walsh

Frequency Dependence Arguments for the Co-evolution of Genes and Culture 211
Graciela Kuechle and Diego Rios

Taking Biology Seriously: Neo-Darwinism and Its Many Challenges 225
Davide Vecchi

Implications of Recent Advances in the Understanding of Heritability for Neo-Darwinian Orthodoxy 249
Martin H. Brinkworth, David Miller, and David Iles

Index 255

Contributors

Sarah Ashelford Division of Nursing, School of Health Studies, University of Bradford, BD7 1DP Bradford, UK, S.L.Ashelford@Bradford.ac.uk

Steven Bond Department of Philosophy, Mary Immaculate College, University of Limerick, Limerick, Ireland, Steven.Bond@mic.ul.ie

Martin Brinkworth School of Medical Sciences, University of Bradford, Great Horton Road, Bradford BD7 1DP, UK, M.H.Brinkworth@Bradford.ac.uk

Eugene Earnshaw-Whyte PhD Candidate, Institute for the History and Philosophy of Science and Technology, University of Toronto, Ontario, ON, Canada, malefax@rogers.com

Gerald M. Edelman The Neurosciences Institute, 10640 John Jay Hopkins Drive, San Diego, CA 92121, USA, eshelman@nsi.edu

Colin A. Hendrie Institute of Psychological Sciences, The University of Leeds, Leeds LS2 9JT, UK, c.a.hendrie@leeds.ac.uk

David Iles Faculty of Biological Sciences, University of Leeds, Leeds, UK

Steve Jones Galton Laboratory, University College London, London WC1E 6BT, UK, j.a.jones@ucl.ac.uk

Graciela Kuechle Witten Herdecke University, Witten, Germany

Rob Lawlor Inter-Disciplinary Ethics Applied, University of Leeds, LS2 9JT Leeds, UK, r.s.lawlor@leeds.ac.uk

David Miller Reproduction and Early Development Group, Leeds Institute of Genetics, Health and Therapeutics, University of Leeds, Leeds, UK

Gonzalo Munevar Humanities and Social Sciences, Lawrence Technological University, 21000 W. Ten Mile Road, Southfield, MI 48075, USA, munevar@ltu.edu

Robert Nola Department of Philosophy, University of Auckland, Auckland, New Zealand, r.nola@auckland.ac.nz

Anthony O'Hear Department of Education, University of Buckingham, Buckingham MK18 1EG, UK, anthony.ohear@buckingham.ac.uk

Alasdair R. Pickles Institute of Membranes and Systems Biology, The University of Leeds, Leeds LS2 9JT, UK, a.r.pickles@leeds.ac.uk

Diego Rios Witten Herdecke University, Witten, Germany, diego.martin.rios@gmail.com

Michael Ruse Department of Philosophy, Florida State University, Tallahassee, FL, USA, mruse@fsu.edu

Timothy Taylor Department of Archaeology, University of Bradford, BD7 1DP Bradford, UK, timtaylor@gmail.com

Davide Vecchi Philosophy Department, Universidad de Santiago de Chile, Santiago, Chile, davide.s.vecchi@gmail.com

D.M. Walsh University of Toronto, Toronto, ON, Canada, denis.walsh@utoronto.ca

Friedel Weinert Division of Humanities, University of Bradford, Great Horton Road, Bradford BD7 1DP, UK, f.weinert@brad.ac.uk

Introduction

Martin Brinkworth and Friedel Weinert

This book provides a forum for the investigation of challenging and open problems at the forefront of modern Darwinism, including philosophical aspects. Modern Darwinism we understand as the synthesis between Darwin's original ideas on evolution and natural selection with the discovery of genetic inheritance. It is the synthesis of the principles of traditional Darwinism, epitomized in Darwin's *Origin of Species* (1859) with the discoveries of genetics, which had their origin in Mendel's laws of inheritance (1865). Although Darwin suspected that inheritance had something to do with disturbances of the reproductive system [1, pp. 131–132] he did not anticipate the principle of genetic inheritance. According to the modern synthesis, genetic changes are random but evolution proceeds in a non-random, cumulative fashion in that it tends to preserve favourable mutations. The neo-Darwinian synthesis was able to insert more detailed explanatory patterns into the existing Darwinian explanations. Whilst the traditional Darwinian explanations appealed to 'descent with modification', the integration of genetics into the Darwinian paradigm, enabled biologists to make much more specific claims about the genetic pathways of this descent with modification.

Kitcher divides the emergence of the modern synthesis into two stages [2, Ch. 3.7]:

I. The mathematical work (by R. A. Fisher, J. B. S. Haldane and S. Wright) brought about the definitive alignment of Darwin and Mendel, and the elaboration of theoretical population genetics, which studies the genetic distribution of

M. Brinkworth (✉)
School of Medical Sciences, University of Bradford, Great Horton Road, Bradford BD7 1 DP, UK
e-mail: M.H.Brinkworth@bradford.ac.uk

F. Weinert
Division of Humanities, University of Bradford, Great Horton Road, Bradford BD7 1 DP, UK
e-mail: f.weinert@bradford.ac.uk

M. Brinkworth and F. Weinert (eds.), *Evolution 2.0*, The Frontiers Collection,
DOI 10.1007/978-3-642-20496-8_1, © Springer-Verlag Berlin Heidelberg 2012

characteristics within a population over a period of time. Neo-Darwinism inserts genetic trajectories into the Darwinian argument, and thereby enlarges the scope of evolutionary explanations. 'In principle, we start from the presence of underlying genetic variation in a population, provide an analysis of the factors that modify frequencies of genes and of allelic combinations, and use GENETIC TRAJECTORIES to derive conclusions about subsequent genetic variation (from which we can arrive at claims about the distributions of phenotypic properties.)' [2, p. 46].

II. Once the mathematical details had been established, connections needed to be forged between the mathematical work and the study of evolution in nature. According to Kitcher, Th. Dobzhansky's 'central endeavour was to articulate Darwin's branching conception of life from the perspective of genetics.' He tried to understand 'how the scheme articulated by the mathematical population geneticists can be instantiated to show how continuous genetic variation has given rise to distinctive local populations that differ in gene frequencies, and how such differences have been amplified over long periods of time to give rise to new species, and ultimately to higher taxa' [2, p. 49]. It is worth noting that Th. Dobzhansky is the author of the famous saying that 'nothing in biology makes sense except in the light of evolution' [3]. E. Mayr's contribution to this phase consisted in analyzing the process of speciation and in providing the 'biological species concept'. Speciation, or the formation of new species, is facilitated by the existence of various reproductive barriers, which prevent members of a species or of similar species, from mating. An important reproductive barrier is geographic isolation of a population, which will lead to the splitting of the lineage from its sister species. Mayr defined a species as a group of 'actually or potentially interbreeding natural populations, which are reproductively isolated from other such groups' [4, p. 120].

Reflecting on the impact of Darwinism, Daniel Dennett used the colourful analogy of Darwin's idea bearing an unmistakable 'likeness to universal acid: it eats through just about every traditional concept, and leaves in its wake a revolutionized world-view, with most of the old landmarks still recognizable, but transformed in fundamental ways' [5, p. 63]. What Dennett had in mind was the monumental transition from a creationist image of the existence of species to an evolutionary explanation. Whilst creationist scenarios highlight the importance of the manifest design of complex organs, like the eye, evolutionary scenarios stress that design is only apparent and can be explained by appeal to natural forces, like selection. The world picture changed from the Great Chain of Being to Darwin's tree of life.

The Great Chain of Being, whose roots lie in Greek thought, provides a hierarchical view of life. The organic world was cast in the image of a ladder, ranking all creatures from the most complex to the most primitive organism, in a descending order of complexity. Each species, including humans, is consigned to a particular rung on the ladder, on which it remains permanently located. There is no room for the evolution of species. The scale of being is static – there is no role for descent with modification [6].

Darwin's tree of life, by contrast, sees every species as descended from some common ancestor with whom it shares some characteristics (homologies). The engine of this branching evolution is the principle of natural selection. Hence the organic world evolves, as a result of random mutations and environmental pressures. Darwin's evolutionary theory is statistical; it holds that there is a tendency in nature to preserve favourable characteristics, while unfavourable characteristics tend to be eliminated. Favourable characteristics tend to aid a member of a species in reproduction and survival.

As it turns out, the questioning of traditional concepts does not stop at modern Darwinism itself. Whilst there is agreement amongst biologists about the fundamental principles of neo-Darwinism, the details of the paradigm do not enjoy such universal agreement. Recent debates, for instance, concern the unit of selection: Is it the gene, as Dawkins believes [7], or is it the individual organism, as Darwin himself and many biologists believe [8]? Or do we need a hierarchical theory of selection, which in addition to the gene and the individual organism includes species selection [9]. A further bone of contention is whether gradualism – the slow, imperceptible work of natural selection over vast spans of time – is the correct way of viewing the operation of selection. S. Gould and his collaborator N. Eldredge proposed instead punctuated equilibrium, a theory that accepts 'the geologically abrupt appearance and subsequent extended stasis of species as a fair description of an evolutionary reality' [9, p. 39].

Orthodox Darwinism places heavy emphasis on adaptationism. Another question concerns the role and range of adaptations: to which extent all features of an organism can be explained as an adaption to ecological challenges. Again Gould offered an alternative: structuralism and the importance of internal constraints. 'We must allow that many important (and currently adaptive) traits originated for nonadaptive reasons that cannot be attributed to the direct action of natural selection [9, p. 1248].

In the present volume the extension of Darwinian concepts takes on two forms: (a) Darwinism is an established research programme but, as several authors argue, it requires extensions, which take it even beyond the Modern Synthesis. As these authors argue the Modern Synthesis fails to pay sufficient attention to the capacities of organisms as the engine of adaptive change. Equally neglected are the roles of developmental biology, epigenetic inheritance and the role of learnt traits, all of which provide opportunities for further genetic change. (b) Darwin's research programme can also be extended beyond the life sciences and applied, as several essays demonstrate, to problems in the social sciences and philosophy. There are in fact many areas where evolutionary concepts have been applied. In this volume we cover evolutionary epistemology, evolutionary medicine, evolutionary economics, the philosophy of mind, and memetics. But evolutionary thinking has penetrated even cosmology [10]. What facilitates these applications are the analogies that seem to exist between evolutionary biology proper and areas beyond evolutionary biology. For instance, Darwin held that the principle of natural selection is able to explain the evolution of mental and moral faculties [11]. In the idiom of the nineteenth century, the brain is the organ of the mind [12, Ch. II.5.2.2]. It is then natural to adopt, as G. Edelman does, an approach to the emergence of the mind from brain functions. Another tempting similarity is the concept of the meme, in analogy with the gene. Finally, analogies of mutation, selection and heredity have been used in models of economic change in a particular industry. However, in all these discussions one should remember the cautionary principle that where 'there is an analogy, there is a disanalogy'. For instance, both genes and memes can be treated as 'replicators' but the meme is a cultural unit, based on the activity of conscious agents, whilst the gene is a biological agent, which follows a blind mechanism. There are also analogies between natural selection and market competition but the aspect of human agency, based on conscious decision, is again a disanalogy.

The tile of the book *Evolution 2.0* is thus a reflection of the ongoing development and extension of the Darwinian paradigm. The essays in this volume present responses to the challenge of how evolutionary concepts can be extended beyond their current realm and what evolution means in the twenty-first century. The open question that exists at the beginning of the twenty-first century is how the details of the neo-Darwinian paradigm will themselves evolve, and what form the 'final' Darwinian paradigm will take. The various chapters reflect the current state of research and thinking in evolutionary biology. They discuss issues that require extensions of the neo-Darwinian paradigm, as well as applications to problems in the social sciences and philosophy. We will introduce these topics now in a little more detail.

Part I discusses Darwinism in Approaches to the Mind. The emergence of consciousness from brain states is still one of the great mysteries of modern science. Darwin tried to deal with this problem in his *Descent of Man* (1871), arguing that consciousness and conscious states may emerge from brain states through the operation of natural selection. The essays in this part deal mainly with modern approaches to this mystery, and some of its implications. G. Edelman discusses a neuro-scientific approach to the mind and consciousness and presents the main ideas of his theory of neural Darwinism, which explains the mind as being 'entailed' by the brain. 'According to Neural Darwinism, the brain is a selectional system, not an instructional one. As such, it contains vast repertoires of neurones and their connections, giving rise to enormous numbers of dynamic states. Behaviour is the result of selection from these diverse states.' As our topic is the human mind, it is natural to ask whether depression, which afflicts the human mind, can be treated from a Darwinian perspective. The chapter "Depression: an evolutionary adaptation organised around the third ventricle" by C. A. Hendrie/A. P. Pickles adopts a strictly evolutionary approach to depression and argues that depression is an evolutionary adaptation, rather than a pathology. S. Ashelford, by contrast, argues that depression should be seen as a maladaptive response to adverse life events. In particular, Ashelford defends a 'separation-distress' system, according to which depression is a developmental problem. Continuing the investigation of mental phenomena, G. Munevar defends a Darwinian approach to the 'Self and Free Will'. He discusses a number of experimental findings, including the results of experiments carried out by his own research team, to argue that the 'self is mostly unconscious', since most of the brain's cognitive functions are also unconscious. One of the controversial issues in recent debates about the mind, in a biological context, has been the notion of the meme. It was popularized by Dawkins, who introduced it as an analogy to the gene concept – both of which he sees as 'replicators.' 'We need a name for the new replicator, a noun that conveys the idea of a unit of cultural transmission, or a unit of imitation. 'Mimeme' comes from a suitable Greek root, but I want a monosyllable that sounds a bit like 'gene'. I hope my classicist friends will forgive me if I abbreviate mimeme to meme. If it is any consolation, it could alternatively be thought of as being related to 'memory', or to the French word même. It should be pronounced to rhyme with 'cream'' [7, p. 192]. The gene is the replicator of genetic evolution, whilst the meme is the replicator of cultural evolution. But in his chapter "The problem of 'Darwinizing' culture (or memes as the new phlogiston)" T. Taylor is highly critical of this notion and sees the 'meme' as more akin to 'the new phlogiston', the imaginary particle, which prior to the discovery of oxygen was employed to explain combustion processes. Taylor argues against the reductionist tendency to 'Darwinize' culture; his alternative view focuses on material technology, which is irreducible to biology. Taylor suggests that 'culture, far from being understandable memically, can be uncontroversially understood as one of those factors extending beyond natural selection that Darwin himself believed also operated.'

Part II deals with the Impact of Darwinism in the Social Sciences and Philosophy. A. O'Hear explains and spells out the limitations of evolutionary epistemology. O'Hear characterizes evolutionary epistemology as the view that 'we reliably know the world (up to a point) because we have been moulded by the world to survive and reproduce in it.' According to O'Hear, evolutionary epistemology fails to explain the full range of activities a self-conscious mind is capable of performing. He therefore proposes, on the basis of the anthropic principle in cosmology a new approach, which he calls Anthropic Epistemology, which holds that 'fine tuning at the beginning of the universe might suggest that life and mind are etched into the fabric of the universe'. The next two chapters discuss aspects of Darwin's methodology, a topic that was close to Darwin's heart. S. Bond considers Darwin's prediction of a moth pollinator for the Madagascar Star Orchid, and asks to which extent a correct prediction can count as evidence in favour of a scientific theory; in this instance evolutionary theory. He compares and contrasts Darwin's prediction with the prediction of the existence of the planet Neptune, on the basis of Newtonian mechanics. Continuing this theme, Nola and Weinert show in their chapter that Darwin, in his work, employed a version of the inference to the best explanation, rather than the hypothetico-deductive method, as has sometimes been claimed. To be precise it is an inference to the most likely explanation: given the available evidence, the inference to the evolutionary hypothesis is more plausible than the inference to intelligent design. In fact, Darwin was clearly aware of the importance of contrastive explanation in science, since the 'long argument', which he develops in the *Origin of Species*, consists in contrasting the explanatory weakness of creationism with the explanatory strength of natural selection. The next two chapters in this part are devoted to a discussion of the application of

Darwinian principles to economics (E. Earnshaw-White) and the ethical treatment of animals (Lawlor). Earnshaw-White discusses an evolutionary theory of economic change (due to Nelson and Winter), which is meant to 'predict and illuminate' economic patterns of growth in a particular industry. He introduces a number of analogies and also disanalogies between economics and evolution, but concludes that the comparison can enhance our understanding of evolution. 'By broadening our gaze from biology, we gain a valuable alternative perspective on evolution by natural selection, which can reconfigure our understanding of evolution in a fashion that may be potentially fruitful from the perspective of biology as well.' Lawlor argues that Darwin's theory can be employed against the position of 'speciesism' in the ethical treatment of animals. Speciesism is the view that discrimination between different animals (in particular humans) and other animals is justified, 'solely on the grounds of species membership (rather than morally relevant considerations such as sentience and self-awareness).' Lawlor poses a number of problems for proponents of speciesism that they need to address, and thereby shows that Darwinism has moral implications.

Part III is concerned with aspects of Darwinism in the life sciences. In particular these chapters deal with various applications and extensions of the neo-Darwinian research programme. S. Jones asks whether 'human evolution is over' and finds, on the basis of genetic information that, at least in the Western world, even if human evolution is not actually over it 'is going very slowly'. An aspect of human existence and evolution is health and illness and M. Ruse introduces the principles of evolutionary medicine, in which natural selection is applied to issues of health and disease. He argues that an appreciation of evolution is crucial in the successful treatment of human ailments. After these illustrations of how to apply Darwinian thinking to the medical field, the next chapters are all concerned with the need to extend the neo-Darwinian synthesis to include further areas. D. Walsh looks at the status of evolutionary theory today and questions whether the twentieth century Modern Synthesis is 'the only reasonable extension' of Darwin's theory. He contrasts the Modern Synthesis with the nascent organism-centred conception of evolution. This alternative accords a 'central explanatory role to the capacities of organisms as the engine of adaptive change'. On the alternative view, an organism's plasticity contributes to the process of adaptive evolution. The next three chapters continue the investigation of how the Darwinian research programme may be extended to include the latest findings in the field. Kuechle and Rios consider the Baldwin effect – a process by which learnt traits become integrated into the genome through a non-Lamarckian mechanism. In order to appreciate the dynamics of 'Baldwinization', the authors argue for a game-theoretic approach rather than positive frequency arguments. D. Vecchi also expresses some dissatisfaction with neo-Darwinism as an adequate paradigm of how evolution works. He argues for a new pluralist theory of evolution, which integrates developmental biology, genomics and microbiology. Finally, M. Brinkworth, D. Miller and D. Iles consider factors affecting inheritance of mutations and epigenetic influences on inheritance. Epigenetics is the study of heritable changes in gene behaviour that do not involve mutation. The latter in particular is identified as having great potential relevance to evolutionary thinking as it may be able to lead to reproductive isolation and hence more rapid rates of evolution within populations sharing the same ecological niche. Returning to one of the themes of the book, they argue that epigenetic inheritance needs to be integrated into current consensus views of neo-Darwinism.

The editors would like to thank the University of Bradford for financial support and in particular our editor, Angela Lahee, at SPRINGER, for her interest in and support of this project.

We are also grateful to Tessa Brinkworth for language editing and creation of the index.

References

1. Darwin, C.: The Origin of Species, vol. 1. John Murray, London (1859)
2. Kitcher, P.: The Advancement of Science. Oxford University Press, Oxford/New York (1993)
3. Dobzhansky, T.: Nothing in biology makes sense except in the light of evolution. Am. Biol. Teach. **35**, 125–129 (1973)
4. Mayr, E.: Systematics and the Origin of Species. Columbia University Press, New York (1942)
5. Dennett, D.: Darwin's Dangerous Idea. Penguin, London (1995)
6. Lovejoy, A.O.: The Great Chain of Being. Harvard University Press, Cambridge, MA/London (1936)
7. Dawkins, R.: The Selfish Gene Oxford. Oxford University Press, Oxford (1989)
8. Mayr, E.: What Evolution Is. Basic Books, New York (2001)
9. Gould, S.J.: The Structure of Evolutionary Theory. Harvard Univerity Press, Cambridge, MA/ London (2002)
10. Smolin, L.: The Life of the Cosmos. Phoenix, London (1998)
11. Darwin, C.: The Descent of Man 1871 (First Edition 1871, quoted from Penguin Edition, Introduction by James Moore and Adrian Desmond). Penguin Books, London (2004)
12. Weinert, F.: Copernicus, Darwin & Freud. Wiley Blackwell, Chichester (2009)

Part I
Darwinism in Approaches to the Mind

The Embodiment of Mind

Gerald M. Edelman

Introduction

The word 'mind' is a loose one with many applications in use. As I use it here, I am restricting it to one definition in *Webster's Third International Dictionary*: 'Mind – the sum total of the conscious states of an individual.' I want to suggest a way of looking at consciousness in tune with, and responsive to, a statement on the subject by the American philosopher Willard van Orman Quine [1]. With his usual ironic candor, Quine said,

> I have been accused of denying consciousness, but I am not conscious of have done so. Consciousness is to me a mystery, and not one to be dismissed. We know what it is like to be conscious, but not how to put it into satisfactory scientific terms. Whatever it precisely may be, consciousness is a state of the body, a state of nerves.
>
> The line I am urging as today's conventional wisdom is not a denial of consciousness. It is often called, with more reason, a repudiation of mind. It is called a repudiation of mind as a second substance, over and above body. It can be described less harshly as an identification of mind with some of the faculties, states, and activities of the body. Mental states and events are a special subclass of the states and events of the human or animal body.

Philosophers have wrestled with the so-called mind-body problem for millennia. Their efforts to explore how consciousness arises were intensified following René Descartes' espousal of dualism. The notion that there are two substances – extended substances (*res extensa*), which are susceptible to physics, and thinking substances (*res cogitans*), which are unavailable to physics – still haunts us. This substance dualism forced confrontation with a key question: how could the mind arise in the material order? Attempts to answer this question have ranged widely. In addition to

This article first appeared in Dædalus (Summer 2006, 23–32), reprinted by permission of the American Academy of Arts and Sciences.

G.M. Edelman (✉)
The Neurosciences Institute, 10640 John Jay Hopkins Drive, San Diego, CA 92121, USA
e-mail: eshelman@nsi.edu

M. Brinkworth and F. Weinert (eds.), *Evolution 2.0*, The Frontiers Collection,
DOI 10.1007/978-3-642-20496-8_2, © Springer-Verlag Berlin Heidelberg 2012

the various forms of dualism, a few proposals we might mention are panpsychism (consciousness inheres in all matter in varying degrees), mind-body identity (the mind is nothing but the operation of neurons in the brain), and, more recently, the proposal that the understanding of quantum gravity will ultimately reveal the nature of consciousness [2]. There are many more proposals, but aside from the extremes of idealism espoused by Bishop Berkeley and Georg Hegel, they all wrestle with one question: how can we explain consciousness in bodily terms?

Attempts to answer this question often begin by examining the features of consciousness to generate a number of more pointed questions. I shall follow that path here. But I don't wish to consider the subject from a philosophical point of view. Rather, I will describe a theory of consciousness based on some significant advances in neuroscience.

Features of Consciousness

Consciousness is a process, not a thing. We experience it as an ongoing series of myriad states, each different but at the same time each unitary. In other words, we do not experience 'just this pencil' or 'just the colour red.' Instead, within a period I have called the remembered present [3], consciousness consists of combinations of external perceptions and various feelings that may include vision, hearing, smell, and other senses such as proprioception, as well as imagery, memory, mood, and emotion. The combinations in which these may participate are usually not fragmented, but instead form a whole 'scene.' Consciousness has the property of intentionality or 'aboutness' – it usually refers to objects, events, images, or ideas, but it doesn't exhaust the characteristics of the objects toward which it is directed. Furthermore, consciousness is qualitative, subjective, and therefore, to a large degree, private. Its details and actual feel are not obviously accessible to others as they are to the conscious individual who has wide-ranging first-person access to ongoing phenomenal experience.

This brief summary prompts me to single out three challenging questions: (1) How can the qualitative features of consciousness be reconciled with the activity of the material body and brain (the qualia question)? (2) Does the conscious process itself have effects? In other words, is the process of consciousness causal (the question of mental causation)? (3) How can conscious activity refer to, or be about, objects, even those that have no existence, such as unicorns (the intentionality question)?

Body, Brain and Environment – The Scientific Approach

There is a voluminous body of philosophical thought that attempts to answer these questions. The efforts of nineteenth century scientists in this regard were relatively sketchy. But a new turn dating from the 1950s has invigorated the scientific approach to consciousness [4]. Neuroscientific investigation has uncovered a rich store of anatomical, physiological, chemical, and behavioural information about our brains. It has become possible to lay the groundwork for a biologically based theory of consciousness, and I believe we are now in a position to reduce Quine's mystery. In this brief essay, I want to lay out some thoughts that bear directly on the nature of consciousness, as well as on how we know, how we discover and create, and how we search for truth. There is nature, and there is human nature. How do they intersect?

In the first place, we must recognize that consciousness is experienced in terms of a triadic relationship among the brain, the body, and the environment. Of course, the brain is the organ we wish to examine. But the brain is embodied, and the body and brain are embedded in the world. They act in the world and are acted upon by it.

We know that the vertebrate species, and specifically in humans, the development of the brain (for instance, the organization of its sensory maps) depends on how our eyes, ears, and limbs receive sensory input from the environment. Change the sequence of actions and inputs to the brains, and the boundaries and response properties of brain maps change, even in adult life. Moreover, we sense our whole body (proprioception) and our limbs (kinaesthesia), as well as our balance (vestibular function), and this tells us *how* we are interacting, consciously or not. We also know that damage to the brain – for example, from strokes involving the cerebral cortex – can radically change how we consciously 'sense' the world and interpret our bodies. Finally, through memory acting in certain sleep states, the brain can give rise to dreams in which our body seems to carry out actions of an unusual kind. The dreams of REM sleep, however fantastic, are in fact conscious states.

Neurology Essential for Consciousness

What can we say about the brain structures whose interactions are responsible for such states? One such interactive structure is the cerebral cortex [5]. Most people are familiar with the cerebral cortex as the wrinkled mantle seen in pictures of the human brain. It is a thin six-layered structure, which, if unfolded, would be about the size of a large table napkin and about as thick. It contains approximately 30 billion neurons or nerve cells, and one million billion synapses connecting them. Moreover, its regions receive inputs from other parts of the brain and send outputs to other portions of the central nervous system such as the spinal cord. There are cortical regions receiving signals from sensory receptors that are functionally segregated for vision, hearing, touch, and smell, for example. There are other cortical regions, more frontally located, which interact mainly with each other and with more posterior regions. There are also regions concerned with movement, for example, the so-called motor cortex.

A key feature of the cortex is that it has many massively parallel nerve fibres connecting its various regions to each other. These cortico-cortical tracts mediate the interactions that are critical for binding and coordinating different cortical activities.

Another structure that is critical for consciousness is the thalamus. This is a relatively small, centrally located collection of so-called nuclei that mediate inputs to, and outputs from, various regions of the cortex. For example, the thalamus processes inputs coming from the eyes via the optic nerves and sends fibres called axons to a posterior cortical region called V1. V1, in turn, sends reciprocal fibres back to the thalamus. Similar thalamo-cortical and cortico-thalamic connections exist for all other senses except for smell; each sense is mediated by a specific thalamic nucleus.

It is known that strokes damaging a cortical area such as V1 lead to blindness. Similar losses of function in other regions can lead to paralysis, loss of speech function (aphasia), and even more bizarre syndromes in which, for example, a patient pays attention only to the right half of his perceptual world (hemineglect). Damage to particular portions of the cortex can thus lead to changes in the contents of consciousness.

The thalamus projects fibres from certain of its nuclei in a diffuse fashion to widespread cortical areas. Damage to these nuclei of the thalamus can have even more devastating effects than cortical strokes, including the complete and permanent loss of consciousness, in what has been called a persistent vegetative state. These thalamic nuclei thus appear to be necessary to set the threshold for the activity of the cortical neurons underlying conscious responses.

The thalamocortical system is essential for the integration of brain action across a widely distributed set of brain regions. It is a highly active and dynamic system – and its complex activity, in stimulating and coordinating dispersed populations of neural groups, has led to its designation as a dynamic core. The dynamic core is essential for consciousness and for conscious learning [6]. Interactions mainly within the core itself lead to integration of signals, but it also has connections to subcortical regions that are critical for nonconscious activities. It is these regions that enable you, for example, to ride a bicycle without conscious attention after having consciously learned how.

The structures I have mentioned thus far function dynamically by strengthening or weakening the synapses that interconnect them. These changes result in the activation of particular pathways after signals are received from the body, the world, and the brain itself. These dynamics allow the development of perceptual categories in the short term and memory in the long term.

In addition to changes that result from and accompany an individual's behaviour, the brain also has inherited value systems selected for and shaped during evolution that constrain particular behaviours. These systems consist of variously located groups of neurons that send ascending axons diffusely into various brain areas. For example, the locus coeruleus consist of several thousand neurons on each side of the brain stem, sending fibres up to the higher brains. Like a leaky garden hose, the fibres release noradrenaline when a salient signal, such as a loud noise, is received. This substance modulates or changes the responses of neurons by changing their thresholds of activity.

Another important value system is known as the dopaminergic system. In situations of reward learning, neurons in this system release dopamine. This compound modulates the response threshold of large numbers of target neurons – for example, those in the cerebral cortex. Without such a value system, the brain would not function efficiently to relate behaviour to the need for survival, i.e. to assure adaptive bodily behaviour. Notice that 'value' as I discuss it here is not 'category'. While value systems constrain rewards or punishments, an individual's behaviour, learning, perception of objects and events, and memory all derive from actions that occur during that individual's lifetime by means of ongoing selections from the brain's vast neuronal repertoires.

A word about the vastness of these repertoires may be in order. Taken together with the intricacy of brain anatomy, the dynamics of synaptic change can give rise to a huge number of possible functional circuits. For example, synaptic change acting on the million billion synapses of the cerebral cortex can provide hyper-astronomical numbers of circuits subject to selection during behaviour.

The Need for a Brain Theory

The background for a theory of consciousness that I have presented so far puts a strong emphasis not just on the action of brain regions but also on their interaction. Some scientists have been tempted to speculate in the opposite direction, claiming that there are 'consciousness neurons' or 'consciousness areas' in the brain. It seems to me more fruitful to ask about the interactions among brain regions that are essential for consciousness.

To explain consciousness in biological terms requires a theory of brain action and a linked theory of consciousness, and both must be framed within an evolutionary perspective. To put these theories in such a perspective, it is useful to distinguish between primary consciousness and higher-order consciousness [3]. Primary consciousness (as seen, for example, in monkeys and dogs) is awareness of the present scene. It has no explicit conscious awareness of being conscious, little or no conscious narrative concept of the past and future, and no explicit awareness of a socially constructed self. Higher-order consciousness, which yields these concepts, depends on primary consciousness, but includes semantic capabilities that are possessed by apes, such as chimpanzees, and, in their highest reaches, by humans who have true language.

To simplify matters, let us focus on the evolutionary emergence of primary consciousness. Why do I insist that we base our explanation on an underlying brain theory? One reason stems from the idea that the neural structures underlying consciousness must integrate an enormous variety of inputs and actions. A parsimonious hypothesis assumes that the mechanism of integration of this great diversity of inputs and outputs is central and not multifarious. A contrasting hypothesis would require separate mechanisms for each conscious state – perception, image, feeling, emotion, etc.

What kind of theory can account for the unity in diversity of these states? I have suggested elsewhere that such a theory must rest on Darwin's idea of population thinking applied to individual vertebrate brains. The resultant theory, Neural Darwinism, or the theory of neuronal group selection (TNGS), states that the brain is a selectional system, unlike an instructional system such as a computer [5]. In a selectional system, a repertoire of diverse elements preexists, and inputs then choose the elements that match those inputs. The enormous diversity in the microscopic anatomy of the brain is created by a selectional rule during the brain's development: neurons that fire together wire together. This rule acts epigenetically, i.e. it does not depend primarily on genes. Overlapping this developmental selection is experiential selection: even after brain anatomy is developed, the connection strengths at the so-called synapses change as the result of an individual's experience. This alters the dynamic signalling across neuronal pathways. By these means, vast – indeed, hyperastronomical – repertoires of circuits, consisting of neuronal groups or populations, are created, upon which further selection can occur and upon which memory is based. As a result, no two brains are identical in their fine details.

The existence of these repertoires is essential as a basis for the selection of circuits leading to behaviour. However, their existence cannot in itself account for the integration of the brain's responses in space and time. For this, a specific anatomically based dynamic feature of higher brains had to evolve. This critical feature is re-entry: the recursive signalling between brain regions and maps across massively parallel arrangements of neural fibres called axons. Re-entrant activity synchronizes and coordinates the activity of the brain regions linked by these axonal fibres. An outstanding example of such parallel connections is the so-called corpus callosum. This tract consists of millions of axons going in both directions to connect the right and left cerebral cortices. Re-entrant activity across such a structure will change with behaviour and also act to integrate and synchronize the dynamic activity of firing neurons. This integrative synchronization allows various brain maps to coordinate their activity by selection. No superordinate or executive area is required. This means that different maps of the brain can be functionally segregated – e.g., for sight, audition, touch, etc. – but, nonetheless, can become integrated, as reflected in the unitary scene of primary consciousness.

What might be useful at this point is an image or metaphor to capture how the re-entrant thalamocortical system – the dynamic core – binds or integrates the complex activities of the various functionally segregated areas of the cortex in a manner consistent with the unitary scenes of primary consciousness. One such image is that of a densely coupled mass of numerous springs. Disturbance within one region of such a structure will be propagated through the whole structure, but certain of its distributed vibrational states will be integrated and favoured over others. Less dense and looser coupling to other springs would correspond to interactions of the core with subcortical brain structures. The main point here is that the myriad interactions in such a densely connected mass will yield certain favoured states, integrating various local changes in a more coherent fashion. This is, of course, only a gross mechanical analogy, but I hope it will help provide a grasp of the subtle electrochemical interactions of core neurons mediated by re-entry that can yield such a great variety of distinct states.

Re-entry is the central organizing principle in selectionistic vertebrate brains. It is of some interest that the underlying structures necessary for dynamic re-entry appear to be missing from insect brains. For our purposes, re-entry will turn out to provide an essential basis for evolutionary emergence of consciousness. The implication is clear: animals lacking wide-scale re-entrant activity are not expected to be conscious as we are.

A Biological Theory of Consciousness

We are now in a position to relate these observations of anatomy and neural dynamics to an analysis of consciousness. As I have suggested, a theory of consciousness based on interactions of the brain, body and environment must be grounded in an evolutionary framework [6]. According to the extended TNGS, primary consciousness first appeared several hundred million years ago at the time of the emergence of birds and mammals from their therapsid reptile ancestors. At these junctures, there appears to have been a large increase in the number and types of thalamic nuclei. Even more to the point, new and massive re-entrant connectivity appeared among cortical regions responsible for perceptual categorization, and more anterior brain regions mediating value-category memory. This is the memory enabled by selective synaptic plasticity, which is constrained overall by value-system responses to reward or to a lack of reward. The integration achieved by this re-entrant system, including the widely distributed thalamic connections, gave rise to unitary conscious or phenomenal experience.

Now we must confront an issue laboured over by students of the mind-body problem. How can one relate the integrated firing of the dynamic core to the subjective experience of qualia? The term ‘qualia’ has been applied narrowly to the warmness of warmth, the greenness of green, etc. In view of the present theory, all conscious experiences – especially the various integrated unitary experiences accompanying core states – are qualia. How can they be explained in neural terms?

The answer harks back to evolution. According to the theory, animals possessing a dynamic core are able to discriminate and distinguish among the myriad interactions of different perceptions, memories, and emotional states [7]. This enormous enhancement of discriminatory capability is of obvious adaptive advantage. Animals lacking a dynamic core can make relatively few discriminations. In contrast, animals possessing primary consciousness can rehearse, plan, and generally increase their chances of survival through their ability to make the vast numbers of discriminations necessary for the planning of behaviour.

This provides a key answer to our question concerning the relationship of neural states to qualia. Qualia *are* the discriminations afforded by the various core states. Thus, although each core state is unitary, reflecting integration of its activity, it changes or differentiates to a new state over fractions of a second, depending on outer and inner circumstances and signals. Still, you might ask: how can we connect neural activity to qualitative experience? The answer is that particular dynamic core states faithfully *entail* particular combinations of discriminations or qualia. Core states do not cause qualia any more than the structure of haemoglobin in your blood causes its characteristic spectrum – the quantum mechanical structure *entails* this spectrum. In this view, conscious states are not causal. The underlying brain and core activity is both causal and faithful. This reconciles the theory with physics – no readjustments for spooky forces need to be made to the laws of thermodynamics to account for consciousness.

What I have not emphasized is the relationship of this model of consciousness to the subjective self. Briefly, this relationship depends on the value systems – the agencies of the brain controlling endocrine and movement responses as well as emotions [7]. In the re-entrant interactions of the core, the earliest and most inherent activities of these systems often supersede other inputs. There is, in foetuses as well as in babies and adults, constant proprioceptive and kinaesthetic input to the core from the body and limbs. It is inevitable that elements of self-reference arise under these circumstances.

This account provides a background for certain features of higher-order consciousness present in humans. With the emergence of higher-order consciousness, through the evolution of larger brains with a new set of re-entrant connections allowing semantic exchange, a socially defined self could appear. Narration of the past and extensive planning of future scenarios became possible. So arose the consciousness of being conscious.

Some find it a retreat to an abhorrent epiphenomenalism to assume that consciousness is not itself causal. But upon reflection, one sees that core processes are faithful ones – so much so that we can speak *as if* our discriminations or qualia are causal. Besides the fidelity of the proposed mechanism, we may point out its universality: all discriminations – whether sensory, abstract, emotional, or fantasy-ridden – are integrated by the same re-entrant mechanisms operating in the thalamocortical core. This lays the burden of differences among qualia on their prior neural origins in regions sending inputs to the core. Qualia are different because the neural receptors and circuits for each differ. Touch receptors and circuits differ from visual receptors and circuits, as do neural circuits governing hormonal and movement responses. Each quale is distinguished by its position within the universe of other qualia, and there is, in general, no place for isolated qualia, except perhaps in the linguistic references of philosophers.

We may now encapsulate the picture put forth here.

According to Neural Darwinism, the brain is a selectional system, not an instructional one. As such, it contains vast repertoires of neurons and their connections, giving rise to enormous numbers of dynamic states. Behaviour is the result of selection from these diverse states. While the brain responds epigenetically to signals from the body and the world, both in development and in behaviour, it also has inherited constraints. These include not only morphological and functional aspects of the body, but also the operation of the brain's value systems.

Such structures and systems were selected during evolutionary time. It is the interplay between evolutionary selection and somatic selection that leads to adaptive behaviour.

To provide for this behaviour, the combinatorial richness and uniqueness of each human brain are coordinated and integrated by the dynamic process of re-entry. Indeed, it was the evolution of new re-entrant circuitry in the dynamic thalamocortical core that allowed the emergence of the myriad discriminations among successive integrated states, which comprise the process of primary consciousness. The rich combinations of qualia constituting phenomenal experience are precisely these discriminations, which are faithfully entailed by core activity. The possession of primary consciousness allows for the planning of behaviour, conferring adaptive advantages on the vertebrate species having this capability.

It is the activity of neuronal groups in the re-entrant dynamic core that is causal, for it provides the means for planning adaptive responses. Consciousness as a phenomenal process cannot be causal in the physical world, which is causally closed to anything but the interactions of matter-energy. Nonetheless, speaking as if conscious states are causal usually mirrors the truly causal core states.

Inasmuch as the set of historic selective events accompanying each individual's development is a function of the unique triadic interactions of body, brain and world, no two selves or sets of brain sets are identical. The privacy and subjectivity of conscious states and selves are an obligate outcome of body-brain interactions. In hominine evolution, a more sophisticated self emerged as a result of social interactions facilitated by the appearance of new re-entrant core circuits that permitted the emergence of higher-order consciousness and, ultimately, language. As powerful as this system of higher-order consciousness is, it still depends critically on the operation of primary consciousness. In any event, the proposed re-entrant core mechanism is universal, i.e., it applies to all mental states, whether they concern emotions or abstract thoughts.

As a result of higher-order consciousness enhanced by language, humans have concepts of the past, the future, and social identity. These enormously important capabilities derive from the activity of the re-entrant dynamic core responding to a multiplicity of inputs from the body and the world, as well as the brain's use of linguistic tokens. The embodiment of mind that results is certainly one of the most remarkable consequences of natural selection.

These considerations provide provisional answers to both the qualia question and the question of mental causation. In this brief compass, I cannot delve deeply into the intentionality question [8]. But the framework I have described posits that consciousness requires re-entry between systems of perceptual categorization and systems of memory. Perceptual systems, by their nature, depend upon interactions between the brain and signals from the body and the world. In one sense they are systems of referral. Moreover, memory systems allow the brain to speak to itself, providing a means for referral to what have been called 'inexistent objects', such as unicorns or zombies. With the emergence of higher-order consciousness and language, intentionality achieves a range that is, for all intents and purposes, limitless.

Significance

I have described a theory, the testing of which will depend on two factors. The first is the self-consistency of its underlying concepts. The second is the provision of support by experimental means. Clearly, it is important to search for neural correlates of conscious processes. There is already evidence that re-entry plays a role in a person's becoming aware of an object [9]. What is required additionally is evidence of how the re-entrant activity of the dynamic core changes when a person goes from an unconscious state to a conscious one. And, of course, we should welcome a variety of experiments exploring neural correlates of consciousness in the hope that some unforeseen correlation will either support or change our theoretical views.

For the present, it is useful to ask what consequences this theory would have, if we assume it is correct. If the theory holds up, we would no longer have to consider dualism, panpsychism, mysterianism, or spooky forces as explanations of our phenomenal experience. We would have a better view of our place in the world order. Indeed, we would finally be able to corroborate Darwin's view that the brain and mind of man are the outcome of natural selection.

Clearly such a theory, linking body, brain, and environment in terms of conscious responses, would, if correct, be of great use in gaining an understanding of psychiatric and neuropsychological syndromes and diseases. Even in the normal sphere, such a theory might give us a better picture of the bases of human illusions, useful and otherwise.

Tangent to these matters, such a brain-based theory might allow us to obtain a clearer understanding of the connection between the objective descriptions of hard science and the subjective, normative issues that arise in ethics and aesthetics. Theory pursued in this fashion might avoid silly reductionism while helping to undo the divorce between science and the humanities.

Quine, with whose quote this essay began, suggested that epistemology, the theory of knowledge, be naturalized by linking it to empirical science, particularly psychology [10]. His proposal encompassed physics, but restricted itself to sensory receptors, a position he justified by claiming that one could, by this restriction, maintain the extensionality of physics. His position, unfortunately, was allied to philosophical behaviourism, and to that extent it skirted the important issue of consciousness. The present excursions, if validated, are more expansive – they would allow the formulation of a biologically based epistemology, which would include the analysis of intentionality. While remaining consistent with physics, this would represent an accounting of knowledge in terms that relate truth to opinion and belief, as well as thought to emotion. Such an accounting would include aspects of brain-based subjectivity in its analysis of human knowledge. Intrinsic to such a study would be the understanding that knowledge, conscious or unconscious, depends on action in the world.

Finally, one must seriously consider the future possibility of an artificial embodiment of mind: we may someday be able to construct a conscious artefact. Brain-based devices capable of acting in the environment and able to develop conditioned responses and autonomously locate targets already exist [11]. Nonetheless, we are still very far from realizing a conscious artefact. To be sure that we had achieved this would require, I believe, that such a device have the ability to report its phenomenal states while we measured its neural and bodily performance. Would such a device sense the world in ways we cannot imagine? Only the receipt of extraterrestrial messages would exceed this enterprise in excitement.

In the meantime, we can take comfort in the fact that such a device, which will not have our body, will neither destroy nor challenge the uniqueness of our phenomenal experience.

References

1. Quine, W.V.: Quiddities: An Intermittently Philosophical Dictionary, pp. 132–133. Belknap Press of Harvard University Press, Cambridge (1987)
2. Penrose, R.: The Emperor's New Mind. Oxford University Press, Oxford (1989)
3. Edelman, G.M.: The Remembered Present: A Biological Theory of Consciousness. Basic Books, New York (1989)
4. Dalton, T.C., Baars, B.J.: Consciousness regained: the scientific restoration of mind and brain. In: Dalton, T.C., Evans, R.B. (eds.) The Life Cycle of Psychological Ideas, pp. 203–247. Kluwer Academic/Plenum Publishers, New York (2004)
5. Edelman, G.M.: Bright Air, Brilliant Fire: On the Matter of the Mind. Basic Books, New York (1992)
6. Edelman, G.M.: Wider Than the Sky: The Phenomenal Gift of Consciousness. Yale University Press, New Haven/London (2004)
7. Damasio, A.R.: The Feeling of What Happens. Harcourt Brace, New York (1999)
8. Searle, J.R.: Consciousness and Language. Cambridge University Press, Cambridge (2002)
9. Srinivasan, R., Russell, D.P., Edelman, G.M., Tononi, G.: Increased synchronization of magnetic responses during conscious perception. J. Neurosci. **19**, 5435–5448 (1999)
10. Quine, W.V.: Ontological Relativity and Other Essays. Columbia University Press, New York (1969), Ch. 3
11. Krichmar, J.L., Edelman, G.M.: Machine psychology: autonomous behaviour, perceptual categorization and conditioning in a brain-based device. Cereb. Cortex **12**, 818–830 (2002)

Depression: An Evolutionary Adaptation Organised Around the Third Ventricle

Colin A. Hendrie and Alasdair R. Pickles

Introduction

Depression is one of the most burdensome diseases in the world [1]. There is some local geographical variation [2] but in general 3–6% of the population are receiving treatment at any one time and there is a 15–25% lifetime chance of becoming depressed [3–7]. Females are universally more vulnerable than men, with 2–3 times more women presenting for treatment [8].

In historical terms the current drug based antidepressant therapies came about in the period immediately following the end of the Second World War and were developed in the rush to find cheap and easily accessible treatments that would, it was hoped, prevent most of the severe social and psychiatric problems seen following the First World War. This search was not well directed by the scientific thinking of the time and so serendipity played an important role [9].

The most widely accepted neurochemical theory of depression is the 'Monoamine Hypothesis'. This holds that depression is caused by a deficit of serotonin and/or noradrenalin in the brain [10–14]. The theory also proposes that the mood elevating actions of the monoamine oxidase inhibitors are due to increased monoamine levels within the synaptic cleft [14] and hence gave rise to the selective serotonin reuptake inhibitors (SSRI's, e.g. Fluoxetine [Prozac]; Paroxetine [Seroxat, Paxil]) commonly used to treat depression today.

Nonetheless, the logic behind the development of these antidepressants is actually no different to the 'Aspirin deficit hypothesis' whereby headache is conceived as being the product of a lack of acetyl salicylic acid because of the therapeutic effects of

C.A. Hendrie (✉)
Institute of Psychological Sciences, The University of Leeds, Leeds LS2 9JT, UK
e-mail: c.a.hendrie@leeds.ac.uk

A.R. Pickles
Institute of Membranes and Systems Biology, The University of Leeds, Leeds LS2 9JT, UK
e-mail: a.r.pickles@leeds.ac.uk

M. Brinkworth and F. Weinert (eds.), *Evolution 2.0*, The Frontiers Collection,
DOI 10.1007/978-3-642-20496-8_3,

administering this compound. Unsurprisingly there is only sparse evidence to support the monoamine hypothesis [15–19] and drugs based on it remain little more than psychiatric Band Aids, in no sense of the word 'a cure'.

In spite of this, patients and medical practitioners have been led to believe that these drugs are highly efficacious (e.g. [17]) whereas the reality is that only about 70–80% of depressed patients respond to drug based antidepressant therapies [15, 19, 20]. Relapse rates are high and the therapeutic lag not only leaves even those potentially suicidal patients who do respond to treatment vulnerable for a period of several weeks, but actually makes them feels worse before therapeutic effects are seen [20, 21]. In consequence, there remains a significant group who prefer to 'self-medicate' with alcohol and/or other non-prescription drugs (e.g. [22]).

Drug based antidepressant therapies are also no longer viewed as the panaceas they once were (e.g. [23, 24]) and perhaps the most damning judgement of all comes from those several pharmaceutical companies that have downgraded their research effort into antidepressant drug discovery, in particular GSK, one of the largest companies in the world who have recently announced that they are pulling out of this area altogether [25]. Against this background, there has been a growing recognition that many of the difficulties in identifying new drug treatments have been produced by the almost universal acceptance of a monoamine hypothesis that is, in fact, wrong [e.g. 17, 19, 20, 26–28].The implications of this realisation for the drug discovery process itself has however been, thus far, less well recognised [29]. The current drug discovery process is glaringly tautological as it involves using drugs that act on the monoamine system as standards in animal based tests that have themselves been developed to be sensitive to and only sensitive to 'known' (i.e. monaminergic) antidepressants. These tests are classified according to their 'predictive', 'face' and 'construct' validities [e.g. 30] on the assumption that there is, as with the questionnaire development that this classification is derived from, some unseen trait that can be uncovered by these models.

Whilst this might be the case under the right circumstances, there is unfortunately, very little consideration given to the species being used and there are no attempts, beyond establishing their sensitivity to monaminergic compounds, made to determine their suitability. In consequence, much of the current literature to do with the development of new antidepressants is based on species that do not actually become depressed [29]. In direct consequence, bizarre procedures, such as suspending a mouse in the air by means of sellotaping its tail to an elevated bar [31, 32] or the placing of laboratory rodents into buckets of water [e.g. 33] have become established as 'models of depression' solely on the basis of their sensitivity to the known (monaminergic) antidepressants.

It should be apparent to even the most casual of observers that these models are not useful in the identification of novel therapeutic entities that work through other, non-monaminergic mechanisms and that it is the tautology of using known antidepressants to define what is and is not a model of depression that prevents the science from progressing and hence determines that a new approach is needed.

The Ethological Approach

No understanding of depression can be complete without full knowledge of its behavioural consequences and the mechanisms that underlie them. Mood state is secondary in this context, as unpleasant as experiencing depressive mood may be.

Examination of the behavioural changes seen in depression [34–37] shows that patients suffer from sleep disturbances and that these frequently involve difficulty in getting to sleep and early morning wakening [e.g. 38]. Appetite is also typically suppressed, as is sexual appetite. Depressives are also lethargic and not motivated to move very much. Partners or relatives of depressed patients notice that they start to neglect their grooming and personal hygiene, have a turned down mouth and a vertical furrowing of the brow. In particular, they adopt a hunched posture and avoid eye contact when being spoken to (e.g. [39–41]).

Taking these behaviours together it is clear that the overall tenor of this cluster is defensive. The hunched posture and avoidance of eye contact are typical human behaviours in response to threat, or the expectation of threat. The decrease in appetite and libido reduces the chances of conflict brought about by competition for these resources and sleep disturbance ensures a period of activity when everyone else is asleep. Social withdrawal places people at the edge of social groups and away from the conflicts that might ensue at their centre. Therefore, in functional terms, depression serves to reduce an individuals' attack provoking stimuli and hence the probability of them being subject to attack by others in their group.

The defensive nature of depression has been recognised by others [e.g. 39, 42] and it has also been suggested that this may have evolutionary origins [e.g. 43]. However, considering depression to be simultaneously an evolutionary adaptation *and* a pathology [43–46] is essentially paradoxical, as selection pressures only work in favour of traits that confer advantage.

In this context, we have recently proposed that depression developed as an evolutionary adaptation that enabled displaced dominant males or females to remain within their social group by facilitating their transition to lower social status [27, 29]. The reproductive opportunities (and/or quality of offspring) at lower social status are usually much poorer (although they must exist to some extent, however poor this is, otherwise there could be no selection pressure towards this adaptation). Hence, our thesis also holds that the most evolutionary significant effect of the loss of social dominance is the loss of the reproductive advantage that came with that status. Therefore, we have proposed that *damage to reproductive potential* is now the key stimulus that triggers depression in those modern humans that have this adaptation, rather than loss of status per se.

This analysis accounts for why it is life events such as the death of a spouse or a child that are major causes of depression [e.g. 47, 48]. It also predicts that females would be most prone to depression because of sexual asymmetry in the costs of human reproduction, whereby females are the most heavily investing sex. This is seen at all levels, from the size of the ova compared to sperm, the energetic costs of gestation and the foreclosed opportunities to attract further mates for a significant period after giving birth [49, 50]. As reproduction has greater costs for females, the costs of damage to reproductive potential (which includes damage to their children, as the carriers of their genes into the next generation) are therefore commensurately greater. This effect is further magnified by the strict time limit imposed on human female reproduction and is predictive of the increase in depression-like symptoms seen during peri-menopause [51]. Relationship breakdown at crucial times in life, loss of a close relative, loss of a job or a history of child abuse all impinge upon an individuals' reproductive potential and this adds further weight to the argument that this is an evolutionary adaptation rather than an illness.

The understanding that depression is an evolutionary adaptation does not however imply that we have to accept its consequences. The psychological pain of depression is no more acceptable to its sufferers than physical pain and the need for more effective treatments is as great as ever. This understanding does however demand that we reassess our views on how depression is mediated and this may be of importance in directing the search for more effective drug-based therapies

The widespread assumption that depression is a pathology has allowed complex explanations to emerge and these have had a tendency to become increasingly more complicated as predictions based on earlier hypotheses were not borne out [e.g. 52–54]. This assumption also focuses attention on only seeking explanations based on pathologies. Further, as stated above, much of the basic work cited in support of such theories has been conducted using laboratory species that are not adapted to become depressed [29] and they are not well supported by the clinical data. For example, in the case of the Corticotrophin Releasing Factor theory, only about half the depressives examined have cortisol levels out of the normal range (e.g. [55, 56]) and predictions based on this theory have not been backed up by community studies [e.g. 57].

Given the lack of progress in improving efficacy of the drug based antidepressant therapies [e.g. 23] there has been a willingness to accept many new antidepressant treatments on the basis of statistically significant but only just noticeable differences (e.g. [58]). They have of course eventually failed and the public have grown cynical.

The application of Occam's razor (the philosophical device that states that where there are two competing explanations for the same set of observations, the simpler of the two should be preferred, all other things being equal) to psychopharmacological science would be timely as it has become the victim of its own lack of rigour (e.g. [59]). Such an approach is known to lead to better predictive theories (e.g. [60]), of the sort that are essential if the continued use of animals in this context is to be justified.

Organisation Around the Third Ventricle

The view that depression is an evolutionary adaptation and not a pathology requires it to have a simple explanation, or at least one that can be traced back to a relatively simple origin. In this context, many of the brain areas mediating the behavioural cluster associated with depression are in close physical proximity to third ventricle. For example, the pineal is involved with the regulation of sleep/wake cycles, the hypothalamus regulates appetite for food and sex and the amygdala, whose main output, the stria terminalis passes through the third ventricle, has an influence on social affiliation as well as fear and defensive behaviours. Hence, it is proposed that this may well be the site where the behavioural expressions of depression are initiated.

Figure 1 shows various anatomical views of the Third Ventricle and the structures that directly border or pass through it. Figure 2 gives brief thumbnails of the main functions each of those structures are thought to perform. This figure also lists the symptoms that may be expected to ensue, were each of those structures to be compromised.

Using this analysis it becomes clear that if the third ventricle is indeed the site of origin for the behavioural cluster associated with depression, the simplest way to

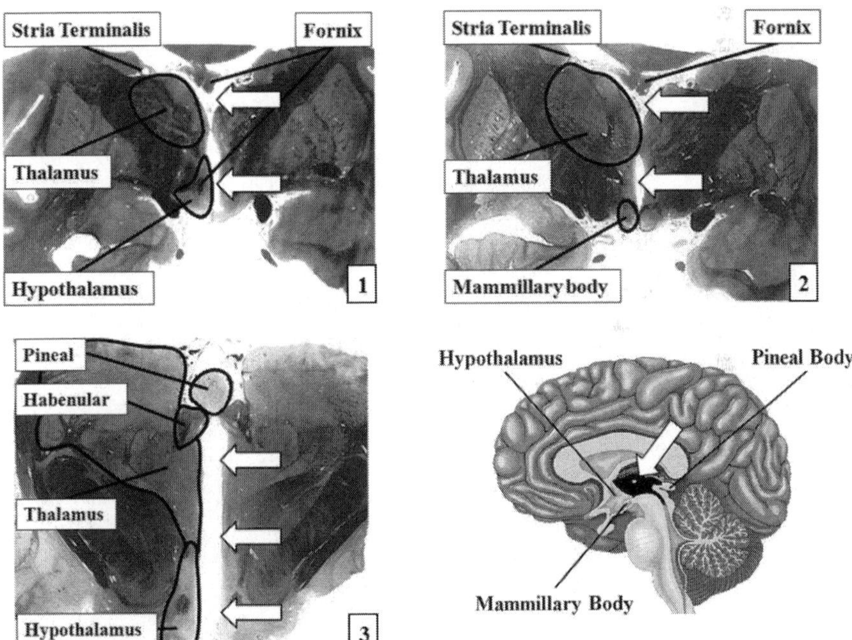

Fig. 1 Detailed view of structures around the third ventricle. The Third ventricle is indicated by the *white arrows* in all panels. *Panel 1* shows a coronal section of Human brain at the region marked 'Hypothalamus' on the Sagittal section (*panel on the bottom right*). The Fornix connects to the Hypothalamus and Mammillary bodies. The Stria Terminalis is the major projection from the Amygdala. *Panel 2* shows the same section at the region marked "Mammillary Body". *Panel 3* is a transverse section across the ventricle. Damage to the pineal and hypothalamus would disrupt circadian rhythms and consummatory behaviour. Damage to the Stria Terminalis would impact upon the Amygdala and so influence fearful and defensive behaviours. Damage to the monoaminergic pathways passing through the Habenular and the effects produced by this, may be incidental.

Directly connected to the Third Ventricle	Main Function	Main Symptom Produced by Disruption of that Function
Pineal Body	Circadian Rhythms (e.g.72)	Sleep disturbance
Thalamus	Integration/modulation	Memory and cognitive deficits
Stria medullaris thalami	Direct connection between Hypothalamus and Pineal	Sleep disturbance
Hypothalamus	Circadian Rhythms Consummatory behaviours	Sleep disturbance, Loss of Appetite, Loss of Libido, Weight loss
Mammillary Bodies	Recall Memory [e.g. 74]	Social withdrawal
Connected via structures that pass through it		
Hippocampus via the Fornix	Complex memory, motivation, emotion [eg.74]	Lethargy, Reduced attention span
Amygdala via the Stria Terminalis	Social Affiliation, Fear/defence Memory for emotional events [eg.75]	Social Withdrawal, Increased defence (hunched posture, averted gaze)
Monaminergic pathways via the Habenular		Events related to non-specific disruption of ascending and descending monoaminergic pathways [eg.76]

Fig. 2 Functions of structures closely associated with the Third ventricle. The table shows those structures that directly border the Third ventricle and those that are connected to it via structures that pass through. The proximal function of the behavioural cluster associated with depression is to reduce the emission of attack provoking stimuli. This is achieved through increased defensiveness, decreased motivation to engage in consummatory behaviours and sleep disturbance, so that there are activity peaks at times when other individuals are inactive. The damage produced by the release of a noxious substance into the CSF is not however precise. Therefore, not every effect produced in this way is important or useful.

achieve this result would be the release of a toxin (or substance that inhibits protection from toxins already present) into the ventricular space. A short burst or pulsatile bursts would ensure the rapid expression of the required behavioural cluster.

Examination of the brains of depressives using post-mortem or MRI techniques reveals a pattern of damage consistent with this proposal. That is, (i) enlargement of the third ventricle, indicating shrinkage of the tissue surrounding it [61]; (ii) evidence for volumetric changes in structures that directly border the third ventricle, such as the mammillary bodies (e.g. [62]) and thalamus (e.g. [63, 64] and (iii) changes in formations that are connected to structures that pass through the third ventricle, such as the hippocampus (via the fornix) (e.g. [65–67]) and the amygdala [68, 69]), via the strai terminalis. Taken together, these observations are compelling evidence that the source of these structural changes emanate from the third ventricle, or at least that this is involved in its transmission.

It must also be predicted that there will be further sites of damage in regions that serve no useful purpose in the expression of the cluster of behaviours associated with depression that has occurred simply because the effects of noxious substances released into the CSF cannot be precisely controlled. The effects on the monoamine systems that pass through the third ventricle via the habenular could well fall into that category.

With regard to the development of new antidepressant therapies, there would appear to be two independent strategies. The first is to seek to identify the proposed noxious substance that is causing the damage with a view to blocking its action or preventing its' release. The second approach is to develop treatments that target the damage and so reverse its effects [27]. In both cases it is essential that this is examined in appropriate species [29] and situations (e.g. [70, 71]) and not in animal models that are just screens of monoamine activity.

In summary, the current manuscript outlines the proposal that depression is an evolutionary adaptation anatomically organised around the third ventricle and that the behavioural expression of this adaptation is mediated by the release of an as yet to be identified noxious factor into the ventricular space. As this is not a precise delivery system, structures beyond those essential for the expression of the depressive syndrome per se will inevitably also be influenced. It is hoped that this analysis will be of heuristic value in the search for more effective drug-based antidepressant therapies.

References

1. World Health Organization: The ICD-10 Classification of Mental and Behavioural Disorders: Diagnostic Criteria for Research. WHO, Geneva (1993)
2. Weissman, M.M., Bland, R.C., Caninio, G.F., Faravelli, C., Greenwald, S., Hwu, H.G., Joyce, P.R., Karam, E.G., Lee, C.K., Lellouch, J., Lepine, J.P., Newman, S.C., Rubiostipec, M., Wells, J.E., Wickramaratne, P.J., Wittchen, H.U., Yek, E.K.: Cross-national epidemiology of major depression and bipolar disorder. J. Am. Med. Assoc. **276**(4), 293–299 (1996)
3. Angst, J.: The epidemiology of depressive disorders. Eur. Neuropsychopharmacol. **5**, 95–98 (1995)
4. Blazer, D.G.: Mood disorders: epidemiology. In: Sadock, B.J., Sadock, V.A. (eds.) Comprehensive Textbook of Psychiatry, pp. 1298–1308. Lippincott, Williams & Wilkins, New York (2000)
5. Keller, M.B.: Depression: a long term illness. Br. J. Psychiatry **165**(s26), 9–15 (1994)
6. Kessler, R.C., Berglund, P., Demler, O., Jin, R., Koretz, D., Merikangas, K.R., Wang, P.S.: The epidemiology of major depressive disorder: results from the National Comorbidity Survey Replication (NCS-R). J. Am. Med. Assoc. **289**, 3095–3105 (2003)
7. Wittchen, H.U., Knauper, B., Kessler, R.C.: Lifetime risk of depression. Br. J. Psychiatry **165**, 16–22 (1994)
8. Burt, V.K., Stein, K.: Epidemiology of depression throughout the female life cycle. J. Clin. Psychiatry **63**(Suppl 7), 9–15 (2002)
9. Bloch, R.G., Dooneief, A.S., Buchberg, A.S., Spellman, S.: The clinical effects of isoniazid and iproniazid in the treatment of pulmonary tuberculosis. Ann. Intern. Med. **40**, 881–900 (1954)

10. Brodie, B.B., Shore, P.A.: A concept for a role of serotonin and norepinephrine as chemical mediators in the brain. Ann. NY Acad. Sci. **66**, 631–642 (1957)
11. Bunney, W.E., Davis, L.M.: Norepinephrine in depressive reactions. Arch. Gen. Psychiatry **13**, 483–494 (1965)
12. Carlsson, A.: Brain monoamines and psychotropic drugs. Neuropsychopharmacology **2**, 417 (1961)
13. Maas, J.W.: Biogenic amines and depression. Biochemical and pharmacological separation of two types of depression. Arch. Gen. Psychiatry **32**, 1357–1361 (1975)
14. Schildkraut, J.J.: The catecholamine hypothesis of affective disorders: a review of supporting evidence. Am. J. Psychiatry **122**, 609–622 (1965)
15. Brown, S.L., Steinberg, R.L., van Praag, H.M.: The pathogenesis of depression: reconsideration of neurotransmitter data. In: den Boer, J.A., Sitsen, J.M.A. (eds.) Handbook of Depression and Anxiety: A Biological Approach, pp. 317–347. Marcel Dekker, New York (1994)
16. Heninger, G., Delgado, P., Charney, D.: The revised monoamine theory of depression: a modulatory role for monoamines, based on new findings from monoamine depletion experiments in humans. Pharmacopsychiatry **29**, 2–11 (1996)
17. Lacasse, J.R., Leo, J.: Serotonin and depression: a disconnect between the advertisements and the scientific literature. PLoS Med. **2**(12), e392 (2005)
18. Mendels, J., Stinnett, J., Burns, D., Frazer, A.: Amine precursors and depression. Arch. Gen. Psychiatry **32**, 22–30 (1975)
19. Stahl, S.M.: Essential Psychopharmacology: Neuroscientific Basis and Practical Application. Cambridge University Press, Cambridge (2000)
20. Moller, H.J., Volz, H.P.: Drug treatment of depression in the 1990s. An overview of achievements and future possibilities. Drugs **52**, 625–638 (1996)
21. Jick, H., Kaye, J.A., Jick, S.S.: Antidepressants and the risk of suicidal behaviours. JAMA **292**, 338–343 (2004)
22. Hendrie, C.A., Sarailly, J.: Evidence to suggest that self-medication with alcohol is not an effective treatment for the control of depression. J. Psychopharmacol. **12**, 112 (1998)
23. Hughes, S., Cohen, D.: A systematic review of long-term studies of drug treated and non-drug treated depression. J. Affect. Disord. **18**, 9–18 (2209)
24. NICE: Depression: Management of Depression in Primary and Secondary Care. Clinical Guideline 23. National Institute for Clinical Excellence, London (2004)
25. Ruddick, G.: GSk seeks to abandon 'White pill and Western markets' strategy, (The Daily Telegraph 5 Feb 2010)
26. Cocchi, M., Tonello, L., Lercker, G.: Platelet stearic acid in different population groups: biochemical and functional hypothesis. Nutr. Clín. Diet Hosp. **29**, 34–45 (2009)
27. Hendrie, C.A., Pickles, A.R.: Depression as an evolutionary adaptation: anatomical organisation around the third ventricle medical hypotheses. Med. Hypotheses **74**, 735–740 (2010)
28. Moncrieff, J.: Are antidepressants as effective as claimed? No, they are not effective at all. Can. J. Psychiatry **52**, 96–97 (2007)
29. Hendrie, C.A., Pickles, A.R.: Depression as an evolutionary adaptation: implications for the development of preclinical models. Med. Hypotheses **72**, 342–347 (2009)
30. Willner, P., Mitchell, P.J.: The validity of animal models of predisposition to depression. Behav. Pharmacol. **13**, 169–188 (2002)
31. Steru, L., Chermat, R., Thierry, B., Simon, P.: The tail suspension test: a new method for screening antidepressants in mice. Psychopharmacology **85**, 367–370 (1985)
32. Cryan, J.F., Mombereau, C., Vassout, A.: The tail suspension test as a model for assessing antidepressant activity: review of pharmacological and genetic studies in mice. Neurosci. Biobehav. Rev. **29**, 571–625 (2005)
33. Porsolt, R.D., Bertin, A., Jalfre, M.: Behavioral despair in mice: a primary screening test for antidepressants. Arch. Int. Pharmacodyn. Thér. **229**(2), 327–336 (1977)
34. American Psychiatric Association: Diagnostic and Statistical Manual of Mental Disorders, 4th edn. American Psychiatric Association, Washington (1994)

35. Beck, A.T., Ward, C.H., Mendelson, M., Mock, J., Erbaugh, J.: An inventory for measuring depression. Arch. Gen. Psychiatry **4**, 561–571 (1961)
36. Hamilton, M.: A rating scale for depression. J. Neurol. Neurosurg. Psychiatry **23**, 56–62 (1960)
37. World Health Organization: The World Health Report 2002: Reducing Risks, Promoting Healthy Life. World Health Organization Geneva, Switzerland (2002)
38. Neylan, T.C.: Treatment of sleep disturbances in depressed patients. J. Clin. Psychiatry **56**, 56–61 (1995)
39. Dixon, A.K., Frisch, H.U.: Animal models and ethological strategies for early drug-testing in humans. Neurosci. Biobehav. Rev. **23**(2), 345–358 (1998)
40. Fossi, L., Faravlli, C., Paoli, M.: The ethological approach to assessment of depressive disorders. J. Nerv. Ment. Dis. **172**, 332–340 (1984)
41. Schelde, J.T.M.: Major depression: behavioral markers of depression and recovery. J. Nerv. Ment. Dis. **186**, 133–140 (1998)
42. Dixon, A.K.: Ethological aspects of psychiatry. Schweiz. Arch. Neurol. Psychiatr. **137**(5), 151–163 (1986)
43. Price, J., Sloman, L., Gardner, R., Gilbert, P., Rohde, P.: The social competition hypothesis of depression. Br. J. Psychiatry **164**, 309–315 (1998)
44. Gilbert, P., Allan, S.: S: The role of defeat and entrapment (arrested flight) in depression: an exploration of an evolutionary view. Psychol. Med. **28**, 585–598 (1998)
45. Sloman, L., Gilbert, P. (eds.): Subordination and Defeat: An Evolutionary Approach to Mood Disorders. Lawrence Erlbaum, Mahwah (2000)
46. Sharpley, C. F., Bitsika, V.: Is depression "evolutionary" or just "adaptive"? A comment. Depression Research and Treatment 10.1155/2010/631502 (2010)
47. Kreicbergs, U., Valdimarsdóttir, U., Onelöv, E., Henter, J.I., Steineck, G.: Anxiety and depression in parents 4–9 years after the loss of a child owing to a malignancy: a population-based follow-up. Psychol. Med. **34**, 1431–1441 (2004)
48. Zisook, S., Shuchter, S.R.: Depression through the first year after the death of a spouse. Am. J. Psychiatry **148**, 1346–1352 (1991)
49. Buss, D.M.: Sex differences in human mate preferences: evolutionary hypotheses testing in 37 cultures. Behav. Brain Sci. **12**, 1–49 (1989)
50. Emlen, S.T., Oring, L.W.: Ecology, sexual selection and the evolution of mating systems. Science **197**, 215–223 (1977)
51. Usall, J., Pinto-Meza, A., Fernández, A., de Graaf, R., Demyttenaere, K., Alonso, J., de Girolamo, G., Lepine, J.P., Kovess, V., Haro, J.M.: Suicide ideation across reproductive life cycle of women. Results from a European epidemiological study. J. Affect. Disord. **116**, 144–147 (2009)
52. Connor, T.J., Leonard, B.E.: Depression, stress and immunological activation: the role of cytokines in depressive disorders. Life Sci. **62**, 583–606 (1998)
53. Middlemiss, D.N., Price, G.W., Watson, J.M.: Serotonergic targets in depression. Curr. Opin. Pharmacol. **2**, 18–22 (2002)
54. Reul, M.H.M., Holsboer, F.: Corticotropin-releasing factor receptors 1 and 2 in anxiety and depression. Curr. Opin. Pharmacol. **2**, 23–33 (2002)
55. Nemeroff, C.B.: The corticotrophin releasing factor hypothesis of depression: new findings and new directions. Mol. Psychiatry **1**, 336–342 (1996)
56. Arborelius, L., Owens, M.J., Plotsky, P.M., Nemeroff, C.B.: The role of corticotropin-releasing factor in depression and anxiety disorders. J. Endocrinol. **160**, 1–12 (1999)
57. Pl, Strickland, Deakin, J.F.W., Percival, C., Dixon, J., Gater, R.A., Goldberg, D.P.: Biosocial origins of depression in the community: interactions between social adversity, cortisol and serotonin transmission. Br. J. Psychiatry **180**, 168–173 (2002)
58. Kramer, M.S., Cutler, N., Feighner, J., Shrivastava, R., Carman, J., Sramek, J.J., Reines, S.A., Liu, G., Rupniak, N.M.J.: Distinct mechanism for antidepressant activity by blockade of central substance P receptors. Science **281**, 1640–1645 (1998)

59. Hendrie, C.A.: The funding crisis in psychophamacology: an historical perspective. J. Psychopharmacol. **24**, 439–440 (2010)
60. Blumer, A., EhrenFeucht, A., Hauseler, D., Warmuth, M.K.: Occam's razor. Inf. Process. Lett. **24**, 377–388 (1987)
61. Baumann, B., Bornschlegl, C., Krell, D., Bogerts, B.: Changes in CSF spaces differ in endogenous and neurotic depression a planimetric CT scan study. J. Affect. Disord. **45**, 179–188 (1997)
62. Bernstein, H.G., Krause, S., Klix, M., Steiner, J., Dobrowolny, H., Bogerts, B.: Structural changes of mammillary bodies are different in schizophrenia and bipolar disorder. Schizophr. Res. **98**, 7–8 (2008)
63. Dasari, M., Friedman, L., Jesberger, J., Stuve, T.A., Findling, F.L., Swales, T.P., Schulz, S.C.: A magnetic resonance imaging study of thalamic area in adolescent patients with either schizophrenia or bipolar disorder as compared to healthy controls. Psychiatry Res. **91**, 155–162 (1999)
64. Dupont, R.M., Jernigan, T.L., Heindel, W., Butters, N., Shafer, K., Wilson, T., Hesselink, J., Gillin, C.: Magnetic resonance imaging and mood disorders: localization of white matter and other subcortical abnormalities. Arch. Gen. Psychiatry **52**, 747–755 (1995)
65. Sheline, Y.I., Sanghavi, M., Mintun, M.A., Gado, M.: Depression duration but not age predicts hippocampal volume loss in women with recurrent major depression. J. Neurosci. **19**, 5034–5043 (1999)
66. Sheline, Y.I., Gado, M.H., Kraemer, H.C.: Untreated depression and hippocampal volume loss. Am. J. Psychiatry **160**, 1516–1518 (2003)
67. Stockmeier, C.A., Mahajan, G., Konick, L.C., Overholser, J.C., Jurjus, G.J., Meltzer, H.Y., Uylings, H.B.M., Friedman, L., Rajkowska, G.: Cellular changes in the postmortem hippocampus in major depression. Biol. Psychiatry **56**, 640–650 (2004)
68. Sheline, Y.I., Gado, M.H., Price, J.L.: Amygdala core nuclei volumes are decreased in recurrent major depression. Brain Imaging **9**, 2023–2028 (1998)
69. Hastings, R.S., Parsey, R.V., Oquendo, M.A., Arango, V., Mann, J.: Volumetric analysis of the prefrontal cortex, amygdala, and hippocampus in major depression. Neuropsychopharmacology **29**, 952–959 (2004)
70. Hendrie, C.A., Pickles, A.R., Duxon, M.S., Riley, G., Hagan, J.J.: Effects of fluoxetine on social behaviour and plasma corticosterone levels in female mongolian gerbils. Behav. Pharmacol. **14**, 545–550 (2003)
71. van Kampen, M., Kramer, M., Hiemke, C., Flügge, G., Fuchs, E.: The chronic psychosocial stress paradigm in male tree shrews: evaluation of a novel animal model for depressive disorders. Stress **5**, 37–46 (2002)
72. Falcón, J., Besseau, L., Fuentès, M., Sauzet, S., Magnanou, E., Boeuf, G.: Structural and functional evolution of the pineal melatonin system in vertebrates. Ann. NY Acad. Sci. **1163**, 101–111 (2009)
73. Tsivilis, D., Vann, S.D., Denby, C., Roberts, N., Mayes, A.R., Montaldi, D., Aggleton, J.P.: A disproportionate role for the fornix and mammillary bodies in recall versus recognition memory. Nat. Neurosci. **11**, 834–842 (2008)
74. Gray, J.A., McNaughton, N.: Comparison between the behavioural effects of septal and hippocampale lesions: a review. Neurosci. Biobehav. Rev. **7**, 119–188 (1983)
75. Aggleton, J.P.: The Amygdala: A Functional Analysis. Oxford University Press, Oxford (2000)
76. Lecourtier, L., Kelly, P.H.: A conductor hidden in the orchestra? Role of the habenular complex in monoamine transmission and cognition. Neurosci. Biobehav. Rev. **31**, 658–672 (2007)

Does Depression Require an Evolutionary Explanation?

Sarah Ashelford

Introduction: Darwin and Emotions

In this bicentenary celebration of Darwin's birth (2009), it is apt to remember that Darwin was one of the first to examine human emotions from an evolutionary point of view in his 1872 book 'The Expression of Emotions in Man and Animals.' In this book Darwin examined the various human and animal emotions and argued that the facial expressions of human emotions are universal (across all cultures), innate and inherited from our non-human primate ancestors. Contrary to many accounts of human emotions today, Darwin argued that many human emotional expressions (behaviours) are not adaptive in humans, but are vestiges of expressions that were adaptive in our non-human ancestors. An example would be the standing on end of hair ('goose bumps') when frightened, which is of little use in humans but may have been adaptive in our mammalian fur-covered ancestors by making them look larger in the face of a predator or threatening con-specific [1]. These vestiges provided Darwin with the argument for the continuity of species: that we have inherited many emotional behaviours and expressions from our non-human ancestors. This argument was made in particular to counter the view that each species was an independent act of creation: that is to say, would a creator design humans with emotional behaviour that has no function? It is surely more logical to believe these expressions are inherited albeit from non-human ancestors.

Clearly then to provide an evolutionary account of human emotional behaviour it is not necessary to provide an adaptive account. Darwin focussed mainly on the *expressions* of emotions, such as frowning, crying, raising the eyebrows, thereby focussing on the nervous and muscular systems of humans and animals. These

S. Ashelford (✉)
Division of Nursing, School of Health Studies, University of Bradford, BD7 1DP Bradford, UK
e-mail: S.L.Ashelford@Bradford.ac.uk

M. Brinkworth and F. Weinert (eds.), *Evolution 2.0*, The Frontiers Collection,
DOI 10.1007/978-3-642-20496-8_4,

outward expressions of emotions can be conceptualised as 'physiological' aspects of emotions; the continuity of physiology between animals and man is well established. Darwin focussed less on the thoughts and feelings associated with human emotions – what we would today call the psychological aspects of emotions, and whose evolutionary origins are much less clear. Darwin did classify many emotions in humans as either 'exciting,' such as rage, which lead to energetic action, or 'depressing' which do not lead to energetic action. Interestingly, in relation to the evolutionary models discussed below, the emotions of pain, fear and grief, are describe by Darwin as exciting at first, but 'have ultimately caused complete exhaustion' [2].

Does Depression Require an Evolutionary Explanation?

My initial answer to this question is yes, depression *does* require an evolutionary explanation, after all, it is a truth universally acknowledged (at least by evolutionists) that 'Nothing in biology makes sense except in the light of evolution' [3]. If depression and other human emotional states are products of the human brain – and its interaction with the body, then like all biological entities it will have undergone biological evolution. Depression may have evolved by natural selection, if it served a useful purpose in our evolutionary past which conferred a selective advantage. On the other hand, depression may be, to paraphrase Darwin, a vestige of an emotional response which was useful in non-human mammalian ancestors. Alternatively, depression may be a 'side-effect', the unfortunate consequence of selection for a different brain function, such as sadness [4] or 'affect reactivity' – the degree to which a person reacts emotionally to various social interactions [5] – see section "iv" (affect reactivity, below).

In today's society, depression is a serious and potentially life-threatening condition. It is among the leading causes of disability worldwide and is projected in 2020 to be the second leading contributor to the global burden of disease for all ages and both sexes [6]

There has been much recent interest in the possible evolutionary origins of depression, motivated in part, by a desire to explain the apparent paradox of how such a severe and debilitating condition can exist today at such high frequency on a global level [7]

The Nature of Depression

There is still significant debate over the nature of depression, and its aetiology. In the West, depression is classified (along with mania) as a mood disorder, and a form of mental illness. Table 1 gives the ICD classification of a depressive illness.

Table 1 ICD-10: symptoms needed to meet criteria for depressive episode (World Health Organisation (WHO) [8])

A

- Depressed mood
- Loss of interest and enjoyment
- Reduced energy and decreased activity

B

- Reduced concentration
- Reduced self-esteem and confidence
- Ideas of guilt and unworthiness
- Pessimistic thoughts
- Ideas of self harm
- Disturbed sleep
- Diminished appetite

Mild depressive episode: at least 2 of A and at least 2 of B

Moderate depressive episode: at least 2 of A and at least 3 of B

Severe depressive episode: all 3 of A and at least 4 of B

Severity of symptoms and degree of functional impairment also guide classification

The collection of symptoms includes not only mood change, but cognitive impairments, slowness of movements and speech, disruptions in sleep and appetite.

In non-Western cultures, however, depression is experienced less in psychological terms, of low mood and negative thoughts, but more in somatic terms, such as physical pain, tiredness, and dizziness [9]. The expression and experience of depression may vary according to culture but there is evidence that depression is universal. Specifically, syndromes involving loss of pleasure and normal interests, sadness or despair, withdrawal from usual activities and relationships, and loss of energy, often associated with important losses of difficult life experiences can be recognised cross-culturally [9].

Evolutionary Models of Depression

The high prevalence of depression in today's society, and the presence of depression in different cultures at similar prevalence rates (between 4% and 10%) have been given as arguments for depression being an evolutionary adaptation in humans [10]. However, depression is not solely a human phenomenon. 'Depression' also exists in non-human animals. For example, it has been observed that infant chimpanzees and macaques when separated from their mothers show distress behaviour which resembles that of depressed humans. As will be described below, Watt and Panksepp [11], have argued that depression in humans results from a re-activation and subsequent shut-down of a conserved 'separation distress' emotional system whose primary function across mammalian species is to engender social bonding with the mother.

In this section I review some of the main evolutionary models of depression.

(i) Social rank models

Many group-living mammals live in dominance hierarchies in which individuals are ranked. An individual's position within a dominance hierarchy is usually established by aggression and conflict, but once established the hierarchy may reduce conflict and engender stability [1]. It is a well-established observation from ethology, that those individuals suffering defeat (or 'social subordination') in group competition suffer considerable stress and exhibit behaviour that resembles human depression. Price et al. (1987) [12] were one of the first to hypothesise that depression in humans signals submission and a 'no threat' signal to rivals. The fatigue and negative, pessimistic cognitions associated with being depressed were hypothesised to prevent any further conflict. In support of this model, it has been argued that when depressed patients are observed they exhibit undue submissiveness and self-derogation, which is similar (at least outwardly) to the submission displays of group-living non-human animals [13]. The idea of depression evolving as a signal of social subordination continues to find support [14]. These social rank models, however, beg the question of whether modern humans evolved in a dominance hierarchy as is seen with our closest relatives, chimpanzees. Perhaps we are seeing the vestiges of a behavioural response which originated before the evolution of the human mind? Additionally, it is not clear whether 'a depressive response' in subordinate animals is an adaptation or not. It is arguable that the depressive–submissive response may be the adverse effect of the increased stress under which subordinates are placed [15]. This possibility is discussed further in the conclusion, below.

Gilbert [16] in an extension to the social rank model, has described the origins of depression as 'the evolution of powerlessness' whereby the vestiges of our evolutionary inheritance mean that in today's society, those who feel trapped in social situations in which one feels inferior or to have lost power, run the risk of developing depression. That is, the depressive response evolved in humans directly from ranking behaviour. Today, it is well-recognised that depressed people experience feelings of loss, defeat, rejection and abandonment [17]. There is considerable evidence that depression is triggered by adverse life circumstances, especially those events that lead to feelings of entrapment and humiliation [18]. Gilbert's model is interesting because it moves beyond describing depression in terms of a behavioural response – that is as a response signalling submission, to incorporate as causal factors, more of what we know of the psychological aspects of depression – which include feelings of low self-esteem, helplessness, loss and so forth [19]. However, it is much more difficult to determine whether these negative thoughts and feelings have a causal role in depression. They may be the consequences of a low mood, rather than the precipitating cause [20].

'Social ranking' models can be conceived in terms of a 'learned helplessness model' of depression [21]. Animals subjected to stressful situations from which there is no escape, soon give up and exhibit inanimate, passive, and often fearful states. This behaviour, described as learned helplessness, persists for months or years, depending on the animal. Moreover this 'helpless' state has many similarities with human depression. In many cases, those experiencing depression feel unable to control that adverse situation in which they find themselves and experience feelings of humiliation and entrapment [18]. They may have in a sense, developed a form of learned helplessness.

(ii) Bargaining
Depression has also been hypothesised as means of 'bargaining'. In such cases, the sufferer signals a willingness to withdraw from the social group – thereby imposing costs on the group, unless the group invests in the sufferer, an act which would help their recovery and role in society [22]. Depression here informs the sufferer that they are 'experiencing (or has recently experienced) circumstances that were reliably associated with net fitness costs over evolutionary time' and thereby functionally shaping investment decision making. This model has been applied in particular to postpartum depression, where risks to fitness through the loss of the infant would be significant. As depression itself has significant morbidity and in some cases mortality, the benefits of such a strategy would have to out way the potential risks to the individual. In today's society this type of behaviour may seem like covert emotional blackmail, however, it may well have paid off for our ancestors.

(iii) Analytic rumination hypothesis
An "analytical rumination hypothesis" (ARH), focuses on providing an evolutionary explanation for the altered cognitions (thoughts) associated with depression [7]. It is suggested that the ruminations associated with depression could help 'focus the mind' on the social problems which caused the depression, and thereby promote the ability to find suitable solutions. As such, it is argued, this ability to ruminate analytically must have provided a selective advantage in our depressed but presumably human, or somehow 'mindful' ancestors.

However, counter to this is the observation that depressed people have poor concentration, and may be deterred (by health-care professionals) from making life-changing decisions until they have recovered. Additionally, Beck's highly influential cognitive model of depression, places the overtly negative thoughts (and ruminations) of depressed people as the *cause* of depression, or at least as important factors maintaining depression, not a way out of depression [20].

(iv) Affect reactivity
Providing an alternative to the adaptationist models of depression, Nettle [5], argues that a predisposition for depression occurs in those at the extreme end of a population distribution of a personality trait described as 'affect reactivity'. Affect reactivity is described as a measure of a person's level of response to negative events, and may be measured by personality dimensions such as 'neuroticism or negative emotionality'. Variability in affect reactivity conforms to a normal distribution suggesting it is multigenic. The optimum level of affect reactivity may be around the population mean, which itself suggests the consequence of stabilising selection. The possession of 'a fairly

reactive negative affect system' is said to cause people 'to strive hard for what is desirable and to avoid negative outcomes, and this may be associated with increased fitness.' Depression is seen as an unfortunate by-product of selection for a more optimum level of reactivity. This model in effect, describes the possible genetic distribution of a personality trait that may cause susceptibility to depression.

In support of this model, genetic studies indicate that susceptibility to depression is multi-factoral, indicating the existence of many genes of small effect [23]. Twin studies give heritability estimates of around 37% [23]. This means that there is variation in susceptibility to depression, of which 37% can be attributed to differences in genes (genetic variance). Adapted traits have low heritability (often below 0.2) because selection has removed much of the genetic variability [24]. This provides further support that depression is not an adaptive trait in humans. Finally, there is a moderate correlation between the genes that predispose to high levels of neuroticism and those that predispose to major depression [25].

(v) Separation distress

An alternative approach to the evolution of the human mind (including depression) is affective neuroscience [26]. This takes the approach of characterising evolutionary-conserved emotional systems. Seven basic emotional systems have been identified as conserved across all mammalian species: (SEEKING, RAGE, FEAR, PANIC, PLAY, LUST and CARE) of which PANIC, in particular may be linked to depression in humans and other animals. Each emotional system is characterised by specific neurochemical networks, defined neurotransmitters and the emotional tendencies it evokes [26].

The PANIC system mediates social attachment, in particular infant–parent attachment following birth. The PANIC system when activated results in 'distress vocalisations' such as crying, which have been termed separation distress. Distress vocalisations arise when young animals are left alone and are inhibited by the close proximity of a caregiver in both humans and other animals. Arousal of the PANIC circuits is hypothesised as one of the major forces that guide the construction of social bonds, with distress vocalisations enabling re-contact with the parent. The neurotransmitters mediating this neural system include corticotrophic releasing factor (CRF) and glutamate, with endorphins and oxytocin inhibiting the system [11].

Watt and Panksepp [11] postulate that a depressive response to separation after an initial protest response has been selected during the evolution of the mammalian brain. The function of a depressive response would be to curtail the distress vocalisations which would minimise detection by predators; depression is conceived as a 'shut-down' mechanism. After a period of intense vocalisations, it might be energetically favourable to 'regress into a behaviourally inhibited despair phase in order to conserve bodily resources.' It is argued that humans have inherited the shutdown mechanism, which can be re-activated in those with a genetic predisposition, or who have other predisposing factors such as early loss or separation trauma, in response to almost any chronic stressor. This response is very similar to the two-stage primate response to loss-separation in which active protest is followed by passive despair [26]. The pain experienced following separation might be a motivation for re-finding the parent. It is noted that the endorphins, the body's endogenous opiate pain-killers, inhibit this system and are produced once the infant re-finds the parent.

The separation-distress response is mediated in part by CRF which activates the hypothalamic-pituitary-adrenal (HPA) axis leading to the secretion of cortisol from the adrenal cortex. These are important mediators of the stress response. The stress response, in particular, elevated levels of cortisol, has long been implicated as a causal factor in the onset of depression in humans [19].

Early separation of children from parents, and other adverse life events occurring in childhood, can lead to neurochemical changes in the brain which predispose to depression in later life [17]. These early life experiences may include loss of a parent, parental neglect and abuse. It is argued that these early adverse life events activate the PANIC/separation distress systems in the brain. The neurochemical changes found in both animal models and in humans subject to early adversity include alterations in the number of receptors for cortisol and CRF. These changes result in impaired regulation of the HPA axis leading to elevated levels of both CRF and cortisol. This neuroendocrine response is activated in animals subject to stress, including the stress of social subordination, discussed above [15]. Impaired regulation of the HPA axis may result in an increased sensitivity to stress in later life.

In the separation distress model, selection has favoured the 'shut-down' as a means of curtailing the distress vocalisation to minimise detection by predators, and/or to conserve bodily resources. The shut down mechanism represents an evolutionary adaptation; one which we have inherited from our non-human ancestors. The strength of this model is that it integrates key neurochemical pathways with emotional responses, particularly those associated with the early attachment bonds made during infancy. The most important neural correlate is the HPA axis whose activation, it is argued, constitutes the first protest phase of separation-distress. The neurochemical mechanism of the 'shut down' is not known but it may be linked to an increased cytokine production or to activation of the dynorphine-opioid system.

The question remains as to whether a specific 'shut-down' mechanism has been selected for its function in curtailing the separation-distress response, or whether a 'shut-down' is an adverse physiological response to prolonged stress [15]. This is discussed further below.

Conclusion

At the beginning of this article I argued that depression requires an evolutionary explanation. This is because depression is an emotional state, the product of the human brain and that the human brain is the result of millions of years of evolution.

The question is whether depression is an adaptation. That is, whether it had a useful function in our evolutionary past, one which conferred a selective advantage on our ancestors. Alternatively, depression may not be an adaptation but a detrimental response, which may occur as a result of prolonged stress.

A common thread in all the models discussed above is that depression is a reaction to adverse life circumstances, in particular those circumstances in which a person feels powerless, subordinate, and/or helpless. The separation distress model, for example, emphasises the distress experienced by the infant associated with separation from the parent. This response may become activated in adult life, in the face of similar loss or adversity. Another common thread is the idea that depression is a form of submission, or state of helplessness, or neurological shut-down. This 'shut-down', or passive behaviour is seen to have an important function, whether it signals 'no threat', conserves energy or indicates the need for greater investment from others. It is this passive behaviour, or shut-down, it is argued that has been selected for, making depression an evolutionary adaptation. Despite plausible functions for a depressive response, a vulnerability to depression may result from selection for another aspect of personality such as affect reactivity [5].

Whilst it is generally accepted that those who develop depression in today's society have an underlying vulnerability, or predisposition, whose nature may be at least partly genetic [23], much research has indicated that this vulnerability is environmental, and frequently associated with loss or neglect in childhood [20, 27]. In other words, depression is seen increasingly as a developmental problem. This echoes the psychoanalytical models of depression, in which depression is viewed as a failure to mourn due to an ambivalent early relationship (usually with the mother). Losses occurring in later life re-activate the original sadness, grief and feelings of abandonment [28]. The biological basis of this may be the separation distress system as has been argued by Watt and Panksepp [11]. We may have, therefore, inherited an important separation distress system, whose function is in childhood to facilitate mother-child bonding. A vulnerability to depression may result according to how our nature is received – that is according to our early environment and our relationship with our caregiver. As such, I find it more plausible that depression in our ancestors was an unfortunate side-effect of the separation-distress response. At the neurochemical level this would represent a form of homeostatic imbalance, in which continued secretion of the stress hormone cortisol leads to the depletion of biogenic amines (noradrenalin, serotonin and dopamine) which are linked to depression in humans [15]. Such an imbalance may cause physiological shut down, but this shut-down is an adverse side-effect rather than a specific mechanism that has been selected for.

It may be argued, therefore, that, if we are to provide an evolutionary explanation of depression, we will need to consider both our evolutionary inheritance (our phylogeny) and our developmental history (our ontogeny).

References

1. Cartwright, J.: Evolution and Human Behaviour, 2nd edn. Palgrave Macmillan, New York (2008)
2. Darwin, C.: The Expression of the Emotions in Man and Animals. Digireads, Stilwell (2005). 1872
3. Dobzhansky, T.: Nothing in biology makes sense except in the light of evolution. Am. Biol. Teach. **35**, 125–129 (1973)
4. Wolpert, L.: Depression in an evolutionary context. Philos. Ethics Humanit. Med. **3**, 8 (2008)
5. Nettle, J.: Evolutionary origins of depression: a review and reformulation. J. Affect. Disord. **81**, 91–102 (2004)
6. What is depression World Health Organisation. http://www.who.int/mental_health/management/depression/definition/en/ (2010). Accessed 10 Mar 2010
7. Andrews, P., Thomson Jr., J.: The bright side of being blue: depression as an adaptation for analyzing complex problems. Psychol. Rev. **116**(3), 620–654 (2009)
8. World Health Organisation: The ICD-10 Classification of Mental and Behavioural Disorders. WHO, Geneva (1992)
9. Kleinman, A., Good, B. (eds.): Culture and Depression: Studies in the Anthropology and Cross-Cultural Psychiatry of Affect and Disorder. University of California Press, London (1985)
10. Keedwell, P.: How Sadness Survived: The Evolutionary Basis of Depression. Radcliffe Publishing, Oxon (2008)
11. Watt, D., Panksepp, J.: Depression: an evolutionarily conserved mechanism to terminate separation distress? A review of aminergic, peptidergic and neural network perspectives. Neuropsychoanalysis **11**, 7–51 (2009)
12. Price, J., Sloman, L., Gardner, R.P., Gilbert, P., Rohde, P.: The social competition hypothesis of depression. In: Baron-Cohen, S. (ed.) The Maladapted Mind: Classic Readings in Evolutionary Psychopathology. Psychology Press, Hove (1987)
13. Gardner, P., Wilson, D.: Sociophysiology and evolutionary aspects of psychiatry. In: Panksepp, J. (ed.) Textbook of Biological Psychiatry. Wiley-Liss, Hoboken (2004)
14. Hendrie, C., Pickles, A.: Depression as an evolutionary adaptation: anatomical organisation around the third ventricle. Med. Hypotheses **74**(4), 735–740 (2009)
15. Sapolsky, R.: The influence of social hierarchy on primate health. Science **308**, 648–652 (2005)
16. Gilbert, P.: Depression: The Evolution of Powerlessness. Psychology Press, Hove (1992)
17. Beck, A.: The evolution of the cognitive model of depression and its neurobiological correlates. Am. J. Psychiatry **165**, 969–977 (2008)
18. Brown, G.W., Harris, T.: Social Origins of Depression: A Study of Psychiatric Disorder in Women. Free Press, New York (1978)
19. Gelder, M., Harrison, P., Cowan, P.: Shorter Oxford Textbook of Psychiatry, 5th edn. Oxford University Press, Oxford (2006)
20. Beck, A.: Cognitive Therapy and the Emotional Disorder. International Universities Press, New York (1976)
21. Millar, W., Seligman, M.: Depression and learned helplessness in man. J. Abnorm. Psychol. **84**, 223–238 (1975)
22. Hagen, E.: Depression as bargaining: the case postpartum. Evol. Hum. Behav. **23**, 323–336 (2002)
23. Sullivan, P., Neale, M., Kendler, K.: Genetic epidemiology of major depression: review and metaanalysis. Am. J. Psychiatry **157**, 1552–1562 (2000)
24. Hartl, D., Clark, A.: Principles of Population Genetics. Sinauer, Sunderland (1997)

25. Kendler, K., Gatz, M., Gardner, C., Pederson, N.: Personality and major depression: a Swedish longitudinal, population-based twin study. Arch. Gen. Psychiatry **64**, 958–965 (2006)
26. Panksepp, J.: Affective Neuroscience. Oxford University Press, Oxford (1998)
27. Bowlby, J.: Attachment and Loss: Sadness and Depression, vol. 3. Hogarth Press, London (1981)
28. Klein, M.: The Collected Writings of Melanie Klein. Love, Guilt and Reparation: And Other Works 1921–1945", vol. 1. Hogarth Press, London (1998)

A Darwinian Account of Self and Free Will

Gonzalo Munevar

Introduction

One of the most important ways in which biology can impact on society is the transformation of long-standing views of human nature. This is particularly true of neuroscience in a Darwinian context, for it has the potential to bring within the realm of science problems about the human mind traditionally beset by philosophical paradoxes. Sometimes, however, the brush with biology seems to make the problems even more acute. For example, recent theoretical and experimental work in neuroscience has been thought to support the claim that both the self and free will are illusions, a claim that has significant and probably adverse implications for our understanding of the mind and even for our commonsense ideas about human relationships and the criminal justice system.

Nonetheless, I will argue in the first part of this paper that neuroscience, in an evolutionary context, actually supports the opposite conclusion: the brain constitutes the self and determines its own actions (i.e. we have free will). More specifically, I will argue that the scientific objections against these notions mistakenly assume that the self and the will must be conscious. They are thus objections against a Cartesian self that has no place in an account of the brain as a highly distributive system evolved and developed in a Darwinian context.

In the second and more suggestive part of the paper, I will apply the same key Darwinian points to examine critically some attempts by neuro-psychologists to determine the nature of the self by doing brain-imaging studies of self knowledge. As we will see, the Darwinian approach points us in new directions in the psychology and neuroscience of the self.

G. Munevar (✉)
Humanities and Social Sciences, Lawrence Technological University, 21000 W. Ten Mile Road, Southfield, MI 48075, USA
e-mail: munevar@ltu.edu

M. Brinkworth and F. Weinert (eds.), *Evolution 2.0*, The Frontiers Collection,
DOI 10.1007/978-3-642-20496-8_5,

Neuroscience Versus Free Will and the Self

Free will, long the province of philosophy, has become in recent decades a subject for theoretical and experimental work in neuroscience. As is often the case with scientific advance, the results are non-intuitive, as the very title of David Wegner's *The Illusion of Conscious Will* [1] suggests. Wegner's work is prompted by experiments, some going back to the 1960s, in which the subject believes he is causing an event (Nielson), though he is not [2], or adamantly denies that he is causing an event (Walter), though he is [3].

In Nielson's experiment, a subject is asked to point-draw with a pencil a straight line on the interior surface of a box (Fig. 1). The box is placed vertically and the subject cannot see his drawing directly, since that surface is parallel to his body. A mirror placed inside the box at an angle of 45° allows the subject to see his hand drawing the line, and this mirror image serves him as his guide. It turns out, however, that the presumed mirror is just a transparent glass, and the hand he sees is not his but that of another person who is point-drawing a line that deviates to the right. The subject nonetheless believes it is his hand, his line, and tries to

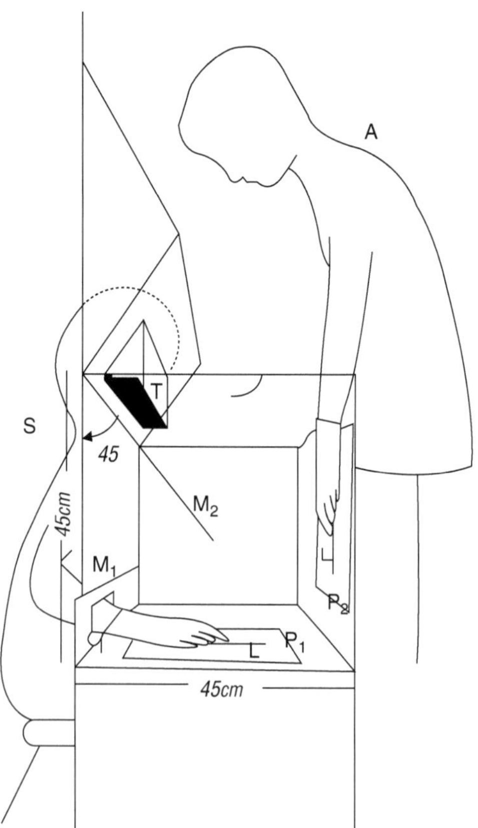

Fig. 1 Subject draws a point line with his left hand, guided by what he thinks is the reflection in the mirror. Unbeknownst to him, the "mirror" is a clear glass and the hand he sees belongs to an investigator that is drawing points off the straight line (From Nielson [2], courtesy of John Wiley & Sons Ltd)

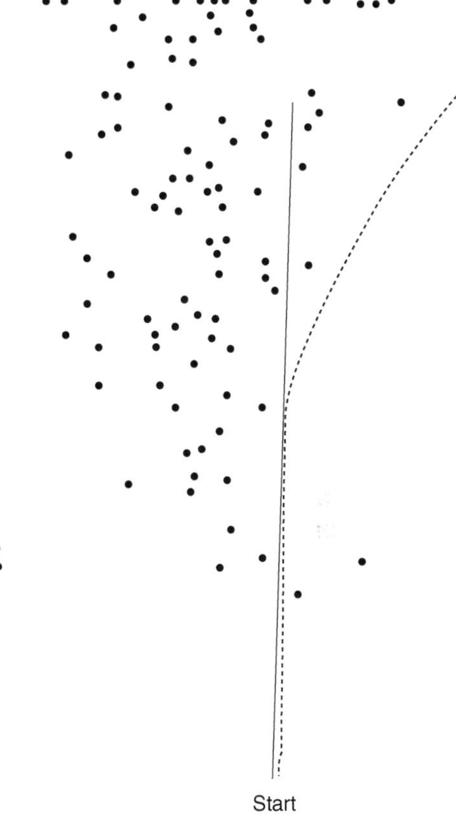

Fig. 2 The subject draws the points to the right, trying to compensate for the points the investigator has drawn to the left of the line (From Nielson [2], courtesy of John Wiley & Sons Ltd)

compensate for the deviation by making points on the paper to the left of the line (Fig. 2). Obviously, he thinks he is responsible for the curve he sees in the "mirror."

W. Grey Walter connected electrodes implanted in the motor cortex of several patients to an amplifier, which in turn sent a signal to a device that operated a slide projector. He then asked the unknowing patients to operate the projector by pressing, at their will, a button that was not really connected to anything. When the patients pressed the button, the experimenter surreptitiously made the projector bring up the next slide. Occasionally, however, the experimenter let the connection of the motor cortex to the amplifier take over. Since electronic transmission is much faster than neural transmission, the projector would bring up the next slide, causing much confusion to the patients, who had not yet pressed the button. Even though they had decided to change the slide, they denied being the cause of the action.

Moreover, willing itself seems to be out of the causal loop, as indicated by a famous experiment in which Benjamin Libet [4] tried to measure how long it took a subject to flex a hand after willing to do so. With a millisecond clock that displayed a large dot instead of hands, Libet timed the conscious thought of willing, the readiness potential as the brain prepares for the motion, and the flexing

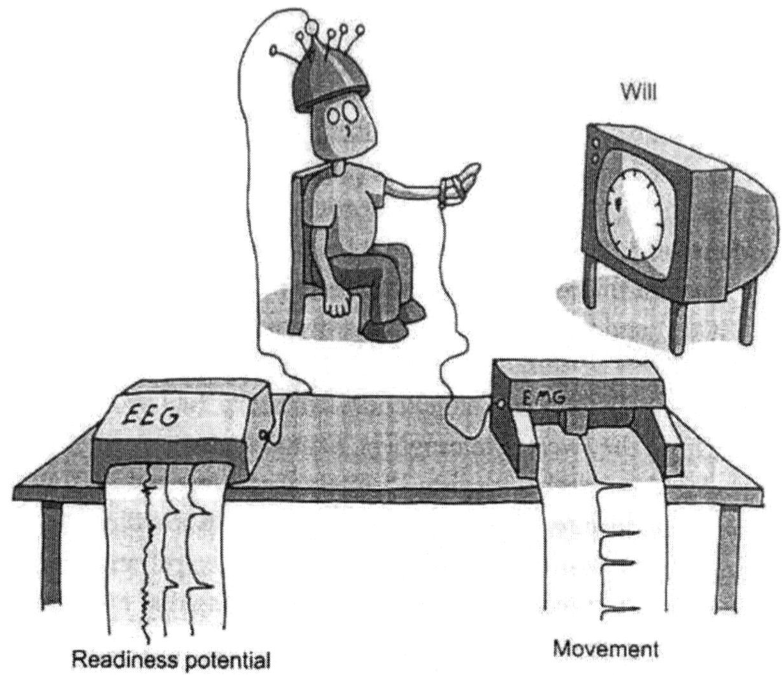

Fig. 3 Subject notices the position of the dot on the millisecond clock at the very moment he wills his hand to flex. An EEG machine determines the moment his brain begins to prepare for the action (readiness potential), whereas the actual motion of the hand is timed by an EMG machine. Libet found that the readiness potential takes place, on the average, 350 ms before the conscious decision to move the hand, which in turns takes place 200 ms before the hand moves (Drawing by Jolyon Troscianko, in Blackmore [5], courtesy of Oxford University Press)

itself (Fig. 3). The timing of the readiness potential was easily measured using an electroencephalogram (EEG), as was the moment the hand flexed using an electromyograph (EMG). To time the conscious event, Libet cleverly asked the subjects to remember the place where the large dot was the moment they decided to flex the hand.

The obvious expectation was that the subject would consciously choose to make his hand move; the brain would give the "order" for the hand to move (readiness potential); and finally the hand would move. Much to most people's surprise, Libet found that the brain began the action (readiness potential) 350 ms, on the average, *before* the conscious "act" of will. The hand flexed 200 ms after that. This suggests that a subconscious event causes both the conscious willing and the flexing of the hand (Fig. 4).

These results seem to settle the old philosophical issue in favor of the position taken by thinkers like Spinoza, Russell and Einstein, all of who thought that free will is an illusion. As we will see, however, all those distinguished thinkers are mistaken: Neuroscience actually supports the notion of free will. But let me begin

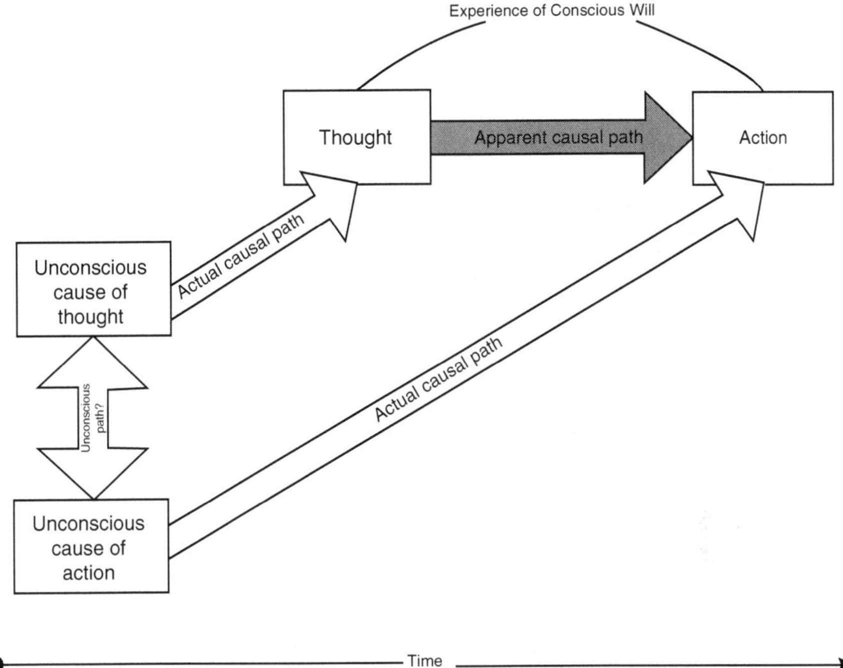

Fig. 4 An unconscious brain process causes the action, and it also causes, directly or indirectly, a conscious thought about the action. Since the thought occurs before the action, we erroneously conclude that the conscious thought is the cause of the action (From Wegner [1], courtesy of The MIT Press)

with a brief philosophical clarification of the problem of free will. A rather common way of thinking about the issue is that the will is a sort of a prime mover inside an agent: an uncaused cause. But as Watson explains [8] the issue should be whether our selves determine our actions. Did *I* determine my actions or did something else (or nothing)? It is not a problem about whether determinism is true, for as Hume pointed out, if my actions are caused by chance they are not mine anymore than if something else did: *I* did not cause them, and thus *I* cannot be responsible for them.

A solution I proposed in previous work [7, 8] was that the brain is the self and the brain/self determines our actions. I will elaborate this solution to some extent below, but it is important to consider first what appears to be a formidable hurdle to any solution of that sort: neuroscience seems to suggest that the self is also an illusion.

A sharp form of the objection is offered by Llinas, who argues [9] that the self is nothing more than an abstraction by which we refer to the most important and generalized cerebral function: the centralization of prediction. There is no centralizing "organ" in the brain, however, no tangible self. The philosopher Daniel Dennett [10] goes even further, for he describes the self as nothing but an abstract center of narrative gravity (of stories that we tell about who we are). Moreover, for

Llinas, the "self" is not fundamentally different from sensory qualities such as colors or sounds, i.e. inventions of the "intrinsic semantics" of the central nervous system. Let me summarize his case against the self: The self centralizes experience, but there is no centralizing organ of experience in the brain, therefore the self is an abstraction, for nothing in the brain corresponds to it. The self abstracts and perception abstracts. These and other considerations lead Llinas to conclude that the self is a form of perception. But perception is also an invention, and thus an illusion. Like most neurobiologists of perception, I will use the term "construction" instead of invention.

A Darwinian approach, however, allows us to meet this objection head on.

A Darwinian and Neuroscientific Defence of the Self

We can pinpoint where these neuroscience critiques of the self and free will go wrong by taking seriously the comparison Llinas makes between the self and perception. In all such critiques, it is assumed that the self, if it exists, should be a Cartesian, conscious self. Likewise, free will should also be conscious free will. This is very odd, however, because neuroscience is the one field that has done the most to undermine the notion of the Cartesian self. We know that most of the brain's functions, including cognitive functions, are unconscious. We should thus *expect* that the self, if it exists, is mostly unconscious. And since free will would be merely the means by which the self determines its own actions, most of our free will should be rooted in unconscious processes as well.

This result is precisely what we find when we think of the brain in the context of evolutionary biology. Any organism needs to demarcate self from other, but in more complex organisms, such as mammals, meeting that need goes beyond the responses of the immune system, for it requires the coordination of external information with information about the internal states of the organism. Such coordination, to be useful, must take into account the previous experience of the organism, as well as its genetic inheritance in the form, for example, of basic emotions that will guide it to survive, reproduce, etc., as Antonio Damasio has argued [11]. Experience must be interpreted on the basis of what the organism takes itself to be, a mostly subconscious task assigned mainly to the central nervous system and particularly to the brain.

A brain that fails to make the connections necessary to carry out this coordinating and interpreting task puts the organism at a disadvantage. It might, for instance, have difficulty learning or remembering crucial facts about its environment, or it might not be able to disambiguate key perceptual information. An interesting example is the case of a patient who suffered from the syndrome of Capgras and insisted that his mother had been replaced by an imposter. V.S. Ramachandran [12] determined that the patient's fusiform gyrus and visual areas were normal and so the patient saw a woman in front of him who looked exactly like his mother. Unfortunately the visual connection to the amygdala had been damaged, and thus the woman in front of him did not *feel* like his mother, hence the Capgras outcome. In contrast, since the equivalent auditory connection had not been damaged, when the patient heard his mother on the phone he had no trouble recognizing her.

A brain that has evolved to unify external and internal information in the context of its own history (or rather its representations of it) is a brain evolved to carry out the functions normally ascribed to a self: being a self is to a large extent what a brain does. Sympathetic views can be found in LeDoux [13] and Monroe [14]. The major mistake that Llinas, Dennett and others make is to confuse the self with its conscious aspects. Another mistake is to make much of the fact that the brain has no homunculus that corresponds to the Cartesian ego. But the brain is capable of carrying out its unification functions precisely because of the extraordinary amount of neuronal connectivity that allows it to synchronize different brain systems at different times, as argued by Edelman and Tononi [15]. A hybrid car has no "motorunculus" whose specific job is to move the wheels. Sometimes it is the electric engine; sometimes it is the gas engine; sometimes it is a combination of the two. However abstract our description of the process, it is always the concrete whole that moves the wheels.

It is true that perceptions can be said to be abstractions in that the brain often "abstracts" from a complicated and complex set of inputs only those features that are relevant, otherwise it would be overwhelmed and unable to respond appropriately [16]. But even if the brain's perception of the self has perplexing phenomenal features without brain counterparts, in particular if such perception is also an abstraction, we need not conclude that there is no such thing as a self. The distinction to keep in mind is that between perceptions (abstracted constructions) and the mechanisms that produce them. An abstract painting may abstract from a scene certain features that the painter chose to emphasize and then place in a different context. The resulting scene is a construction, an abstraction, but the painting itself is a physical object shaped by physical actions. Likewise with the self, whatever the abstractions by which we perceive it consciously.

Illusions and the "Conscious" Self as Perception of the Brain's Self

Llinas argues that the conscious self is as much a construction ("invention") of the brain as secondary sensory qualities are, and so does Dennett. This suggests that the conscious self is a form of perception. But just as a tree and our perception of the tree are different things, so are the self and our perception of the self.

We may be struck by the strong analogies between the illusions of sensory perception and those of consciousness and free will. Indeed the illusions of perception are illusions of consciousness. There are many visual illusions. For example, the brain cannot help but see as three dimensional some drawings in two dimensions, or to see a line as longer than another exactly of the same length because of the visual context. The brain fills in the blind spot, and it also falls prey to change blindness and inattentional blindness. Similar illusions affect the other senses. Properly spaced touches – five on the wrist, three on the elbow and two near the shoulder – makes us feel as if a rabbit where hopping up our arm ("cutaneous rabbit"). All seem paradoxical if we expect that the job of the brain in perception is to give us a "picture" of the world that corresponds to reality point by point. Llinas is not alone in casting doubt upon this realist view, for his approach is shared by a great many cognitive neuroscientists such as Hoffman [17], Johnston [18], Koch [19], and Edelman and Tononi [15]. The brain is the result of evolution and must be seen as shaped, at least in its basic functional structure, by natural selection. A brain's perceptual "constructions," thus, are typical in a species because of their past successes in the interactions between that species' ancestors and their environments, regardless of "correspondence," as I have argued previously [20–22]. If it works well enough, a perceptual mechanism may continue to use strategies that have worked well generally, even if they lead to perplexing experiences from time to time. Thus it should not be surprising that we sometimes have difficulties, say, deciding on the contents of consciousness.

All the problems with the illusions of consciousness in general, and with those of consciousness of the self, are simply what should be expected from all forms of perception. But, once again, when we are conscious of our selves we merely have a convenient and simple way to perceive an aspect of a very complex neural interplay. We only get the "bottom line," so to speak. Hoffman [16] equates perceptions with icons on a computer screen: They are not at all like the computer circuits and processes they stand for, but they are a very practical way of dealing with the task at hand. A brain organized in a certain way constitutes a self. The "conscious self" is thus a perceptual "icon" of some practical value.

This account gives us a consistent picture of the brain (sensory perception, self, etc.) in terms of evolutionary theory.

This is not to deny that the problem of the self is connected with the problem of consciousness, but it is a mistake to suppose that our conscious experiences must be experienced by a conscious "someone," hence the "conscious self." When Llinas, Dennett and others point out that nothing in the brain corresponds to that conscious "someone," they conclude that the self does not exist. But our experiences are indeed experienced by a self that is made up of a large distributed system of brain structures and, though obviously capable of conscious experience, most of its operations are not conscious. Even when some of them are, in Francis Crick's words, we may be aware of a "decision" taken by the brain "but not of the computations that went into that decision," for those computations are not open to consciousness [23]. When a brain tries "to explain to itself why it made a certain choice," sometimes "it may reach the right conclusion. At other times it will either not know, or, more likely, will confabulate, because it has no conscious knowledge of the 'reason' for the choice" [23]. On confabulation see also P.S. Churchland [24], Gazzaniga [25] and Gazzaniga and LeDoux [26]. To explain fully how the brain works, we need to explain the problem of consciousness, simply because to explain fully many of the operations of the brain we need to explain just how consciousness fits in, and how it came to fit in.

Nevertheless, the remarks by Crick and the results of Libet's 1985 experiment, taken in the context of evolutionary biology, indicate that the self and free will are distinct from their conscious aspects.[1]

A Brief Sketch of the Argument for Free Will

In previous work [7] I pointed out, as others have, that the nature of the brain is such that many brain processes are emergent in that the elements (the initial strength of synaptic connections in the primary sensory areas, for example) do not fully determine the whole process. But that work also pointed out that such processes are emergent in a stronger sense, for the elements themselves are determined by the whole (for example, in the selection and top-down modulation of neuronal information when the brain makes sense of an ambiguous perception or complex situation). The result is a non-linear system that, although using the laws of physics and chemistry, adds its own "laws" to transform external "information" into its *world*. This *sui generis* system, the brain as its own self, determines what actions it deems most appropriate. *My* actions are characteristically determined by my *self* as instantiated in my brain.

Although the world exerts an influence on the brain's decisions (e.g. stimuli, genes), a strongly emergent system such as the brain amounts to a pocket of the world ruled by emergent "laws" of its own. That is, there is a discontinuity between the world and the "laws" by which each individual brain interprets a situation, finds it relevant, evaluates, and decides how to deal with it. Physics and chemistry operate in a brain, but it is the dynamic organization of that brain that places the whole of its elements beyond the behavior of mere falling bodies, just as it is the organization of those elements that makes their joint action intelligent. Moreover, neurons are not only recruited at different times for different networks to carry out functions of the brain, but the strength of the synaptic connections are modulated by a complex array of influences by other neurons, as P. M. Churchland points out [27, 28], and they thus interact, combine, and compete with other such "alliances" for central stage, as Edelman and Tononi explain [15]. It is strong emergence in that the contribution of an element of the self to a decision then depends partly on the systemic influences of the self on that element.

Strong emergence can be found in non-biological systems as well. Rayleigh-Benard convection, for example, can be produced when the temperature difference

[1]Although Damasio's work has inspired several of my Darwinian remarks about the self, we part company in that he also integrates consciousness into his multi-level scheme of the self (core consciousness is associated with the core self and his notion of autobiographical self comes with extended consciousness).

between two plates containing a fluid becomes quite large. The result is a series of parallel cylinders, Benard cells, which according to Chemero and Silberstein [29] exhibit properties of "integrity, integration and stability... determined by the dynamic properties of the nonlocal relations of all fluid elements to each other." This is a case in which, not unlike the brain, "[T]he large structure supplies a governing influence constraining the local dynamics of the fluid elements." One difference with the brain is that the latter involves far greater complexity and plasticity.

Concerning free will, as Watson puts it, "the question is how a series of natural processes (for which you are not accountable) can result in processes and events over which you do have control (for which you are accountable)." Given my explanation of strong emergence, having a pistol pointed at one's head, which most would consider an external motive for action, requires that the pistol be recognized as such (not as an illusion or a toy pistol), that other clues be read as confirming the impression of danger, and so on. We may have no control over the pistol, but it does not provide a motive for action until it is processed by our brain and integrated in a certain way. Then we make a decision guided by the characteristics, history and present circumstances of our particular brains: some would run, some would take the bullet for a loved one, some would kick the hand with the pistol, etc. The decision is marked by the idiosyncratic nature of the brain that has control over it. This point holds for character as well, for character is built decision by decision over the course of our lives. Although many external and internal processes will influence those decisions, they cannot do so until they are assimilated into the whole by emergent mechanisms. When the self determines a decision, it does so *qua* self, for it is the self that controls the relevant factors, that assigns to them values within the system, makes them relevant, and compares and combines them with other factors. Unless they are made part of its *sui generis* world, they would play no role in the decisions the self makes.

The self characteristically, then, determines its own actions, and is thus morally responsible. Contrast this conclusion with what happens when factors outside the organic assimilation and control by the brain determine a person's actions: It is clear then that we cannot assign moral responsibility. Suppose we implant some electrodes into Peter's brain. If Peter's self would have decided to stand up, but through radio signals we alter his decision so as to stay seated instead, Peter is not acting freely; even though consciously he only gets the "bottom line" of the neural mechanism, he may still *feel* that he is acting freely. When Alzheimer's robs John from access to his past, when the continuity of his self is disrupted, we are not entitled to assign blame: He is no longer himself. And when a disruption in the proper rate of neuro-transmitters render Mary completely unable to interpret a situation as she would under normal circumstances, or gives extraordinary significance to an event that would not normally be that important to her (as in drug-induced paranoia), we should exempt her (nor merely excuse her) from moral blame to the degree of her inability. Of course we may blame her for her choices that made her into a drug addict, but the reason for this harsh judgment is presumably that her brain was working normally then.

Someone might object, however, that it is not clear how we can hold a person responsible for his "unconscious" decisions, since he could not, by definition, be aware of them or "in control" of them.[2] Nevertheless, if most of our brain activities are unconscious and the conscious event is but the mere conclusion of a myriad neural processes, we do indeed hold people responsible for what is largely unconscious. And we would be right, under this account, because each person's self is "in control" of his or her actions. Now, I have claimed that most of our mental life is unconscious, but this is not to say that all of it is. Some conscious events may be necessary for the brain to assimilate some experiences, for example. If so, consciousness would play a part in the process by which the brain evaluates those situations and decides to act on them. Most likely in those cases, the subject would have a conscious experience of willing a certain action in response, just as Libet's subjects have such experiences even though it is the brain that initiates the action. Although possibly required in some cases of perception and others, however, the conscious experience need not be present in all cases in which we do assign responsibility. Take the simple example of driving a car from our home to our office. We are hardly conscious of the specific decisions, hundreds of them, we take in that drive. Nonetheless, as a citizen-driver you must train yourself to snap to attention whenever anything out of the ordinary takes place during an otherwise forgettable drive to work. If you do not snap to attention, if you do not become aware that a child has run onto the road to retrieve a toy, you may well kill him. Can you then say in your defense, "I was not aware of the child on the road and therefore I am not responsible for his death"? The point is that as a citizen-driver you do have an obligation to train your subconscious to kick your brain into high alert at moments like that. Otherwise you may be considered negligent, and thus blameworthy, even though you never became aware of the child's presence on the road.

The unification and interpretation of experience allows the organism to behave in ways that seek its advantage. It is the brain as a self, then, that determines the organism's actions. But what of the experiments by Nielson and Walter? Did they not show that free will is an illusion?

No. They only show that the so-called "conscious will" is subject to illusions. As I remarked earlier, if consciousness of the self is a sort of internal perception, then we should expect perceptual illusions. Thus, illusions of the conscious will do not imply that the brain does not exercise its will anymore than visual illusions imply that the brain does not see.

Indeed, Wegner [1] has already provided a plausible explanation for why those illusions of the *conscious* will come about. When accounting for Libet's experiment (Fig. 4), Wegner tells us that the brain first begins to plan and bring about the action; second, the brain causes conscious thoughts about carrying out the action; and third, the action takes place. For us to believe that we have caused the action, Wegner explains, three conditions must be fulfilled: (1) the conscious thought occurs before

[2]James S. Rodgers, personal communication.

the action; (2) the thought is "consistent" with the action; and (3) we cannot detect other possible causes of the action. By "consistent with the action" I think he means that the resulting action is roughly of the sort the subject had thought about.

Using Wegner's conditions, let us examine the two troublesome experiments. First, in Nielson's experiment, we realize that the subject has the conscious intention of point-drawing a straight line (1); he sees the points through the false mirror (2), but (3) does not know that it is a trick (he is seeing someone else's hand). Of course, the points are not appearing where he wishes them to appear, but this does not affect condition (2), just as a player who misses a penalty kick blames himself for his poor execution. Thus the subject in Nielson's experiment concludes that he is responsible for a badly drawn line.

Similarly in Grey Walter's experiment, whenever the subjects pressed the button and the slide changed, they mistakenly believed that they had caused the action, for conditions (1) – (3) were fulfilled. But when the slide changed as the direct consequence of the decision by the subject ("taken" out of his brain directly by the electrodes) and before he pressed the button, the subject did not believe that he had been the cause of the change in slide, even though he was. In this case condition (2) was not fulfilled, since his thought was that if he pressed the button, the slide would change; but he had not pressed the button, and he could not observe any other possible cause of the effect. He felt very confused as a result.

These kinds of explanations, we may note, do not seem very different in kind from the kinds of explanations that we may give for perceptual illusions. I presume this point applies to new and future experiments concerning "conscious will."

I will then summarize this argumentative part of the paper as follows:

1. It exposes a mistake in the interpretation of the impact of neuroscience on society, and then corrects it.
2. It determines that the mistake is based on the adherence by Llinas, Dennett and others to the notion that criticizing the concepts of conscious self and conscious will amounts to a criticism of the existence of self and free will. Self and free will are largely subconscious, even if we can access them consciously some of the time.
3. When those concepts are adapted to the advances of neuroscience in the context of evolutionary biology, we see that (a) there are good reasons for thinking of the brain as constituting the self, and (b) there are good reasons for thinking that the self then determines our actions.
4. The last result may be, then, sufficient to conclude that we are morally, and socially, responsible for our own actions, thus solving the problem of free-will.

The Darwinian Self: From Theory to Experiment

This final section is mostly suggestive, since the main evidence discussed is experimental work still in progress. It is offered to show how the Darwinian considerations about the self of the first half may lead to a change in direction at the experimental level in psychology and neuroscience.

Even though I will concentrate on the self, I should mention that my ideas concerning free will should be examined in the context of the experimental results and the theoretical claims made in the contemporary neurobiological studies of voluntary choice and addiction, particularly in work presumably aimed to elucidate the neurobiology of free will. Much of that work deals with drug addiction. For example, in addiction, as Baler and Volkow tell us, "there seem to be intimate relationships between the circuits disrupted by abused drugs and those that underlie self-control. Significant changes can be detected in circuits implicated in reward, motivation and/or drive, salience attribution, inhibitory control and memory consolidation" [30]. Moreover, Volkow and Li argue that drug addiction causes long-lasting changes in the brain that undermine voluntary control [31]. That examination, however, is beyond the scope of this paper, and all I can offer here is a promise to undertake it in the near future.

Now, if consciousness of self is a form of perception, it is then an internal sense. But a sense of what? We have seen above that a brain worth its salt in evolutionary terms must form a self. And Damasio shows that the brain "represents" to itself internal states of the organism in a variety of ways, including what he calls "emotional tone." It is plausible, then, that the brain also gives us ways of reading the functions of the self. In consonance with the Darwinian view described above, we should expect the self to be distributed over many regions of the brain, and we should also expect that conscious access to the self would depend on the activation of different brain regions, given the particular functions of the self targeted. Such expectation seems to be met in the neuroscience of the self.

One area of interest, for example, concerns brain imaging studies of self-knowledge, which have produced a variety of apparently inconsistent results. According to Keenan, self-recognition is correlated with activation of the right prefrontal lobe, which leads him to link self-awareness, or self-knowledge, and thus the self, to that region [32, 33]. This line of thought is based on Gallup's experiments with chimps that recognized themselves on a mirror (as shown by their touching spots painted on their faces) [34, 35], on brain-imaging studies that detected activation of the right prefrontal lobe during self-recognition tasks, and on reports of lesions of the right prefrontal lobe and their adverse effects on self-recognition, as well on the seemingly innocent assumption that self-recognition is a form of self-knowledge. There is more to self-knowledge than self-recognition, however, as Morin points out [36], and, very importantly, other brain-imaging studies of self-recognition seem to emphasize activation of left-hemisphere regions. Of particular importance seems to be the finding by Macrae, Heatherton and Kelley that medial prefrontal regions of the cortex (MPFC) are recruited in self-referential and mentalizing (theory of mind) tasks but remain inactive when referring to other cognitive tasks [37]. Brain-imaging studies that influenced the functional magnetic resonance imaging (fMRI) study by Macrae and colleagues include Craik et al. [38] and Kelley et al. [39]. Other interesting work, such as Platek's fMRI study on the recognition of self in photos [40], "Using personally familiar gender and age matched control faces…found a distributed bilateral network involved in self-face recognition that included right superior frontal gyrus, right inferior parietal lobe, bilateral medial frontal lobe, and left anterior middle temporal gyrus" (but no MPFC) [41].

This pluralism of results is what should be expected, given my previous remarks. Morin agrees [36] as he approvingly quotes from Gillihan and Farah's meta-analysis of brain-imaging and neuropsychological studies of self-knowledge:

> Had the points clustered in certain regions or along certain networks, the hypothesis of a unitary self system would have been supported. However, neither the imaging nor the patient data implicate common brain areas across different aspects of the self. This is not surprising because there is generally little clustering even within specific aspects of the self [42, p. 94].

Most of these authors treat the self and consciousness of the self as one and the same, a conflation against which I argued in the first part of the paper. But insofar as the distinction seems to be made, the conclusion should be the same: the brain regions involved in the perception of the self should also be distributed. Turk et al. [43] seemed to agree: "The available evidence suggests that the sense of self is widely distributed throughout the brain." But several of the collaborators on that paper (Heatherton, Mcrae and Kelley) went on to produce some of the most important work that gives primacy to the MPFC. That work includes an important collaboration with Moran [44]. Also important is Moran's collaboration with Saxe [45].

Moran's fMRI study found that the medial prefrontal cortex was activated by the self-relevance of personality traits, but not by whether those traits were thought to be positive or negative. Saxe et al. add that their "current data provide the strongest evidence to date that sub-regions of medial precuneus and MPFC are recruited both when subjects reason about a character's thoughts, and when they attribute a personality trait to themselves," although in other regions they do not overlap. The sweeping claims about the significance of the MPFC for the self need even stronger evidence, it seems to me. In particular they need an experiment that rules out other cognitive factors as causes of the activation. That would be, for example, an experiment in which brain activation during the self-attribution of traits is compared with the attribution of the same traits to others, preferably in two groups: one of people familiar to the subject, and another of people not familiar. If the MPFC lights up during the self-attribution, but not during the other conditions, we may safely conclude, given the evidence from the other studies, that it is the connection with the self that activates the MPFC. Otherwise it would seem instead that it is the mentalizing aspects of a variety of tasks that cause the activation.

Saxe et al. themselves bring up an important study on self-knowledge by Lou et al. that seems to fit the bill [46]. This study compared regional cerebral blood flood (rCBF) changes as determined by positron-emission tomography (PET) during a retrieval memory task concerning prior judgments about personality traits of Self, Best Friend and a Celebrity (the Danish Queen, since the subjects were Danish). The study found massive activation of medial prefrontral and medial parietal/posterior cingulate regions in the performance of all tasks, consistent with previous studies. But it also found that recalling self-relevant judgments

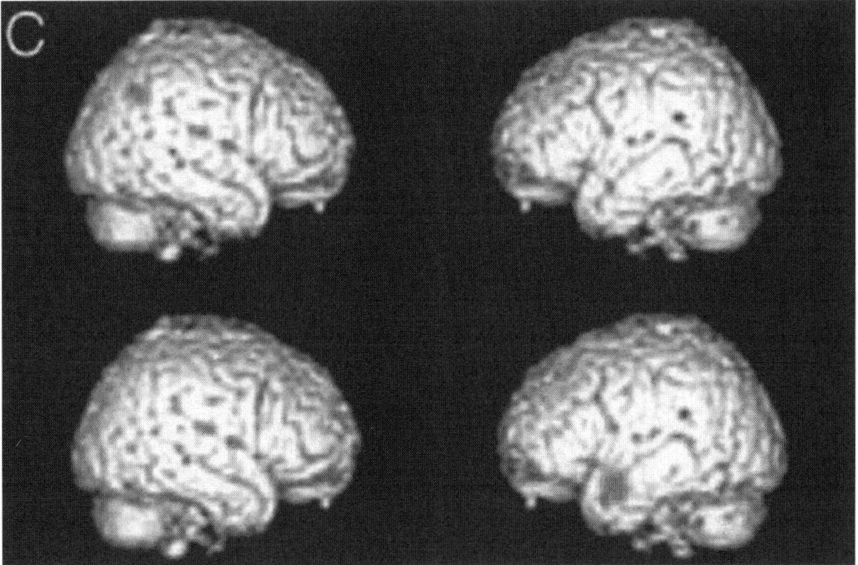

Fig. 5 When subjects recalled whether they attributed a series of personality traits to themselves, the right parietal cerebral cortex was differentially activated; but when they recalled whether they attributed those personality traits to others, the left temporal cortex was differentially activated instead (According to Lou [46]. Copyright (2004) National Academy of Sciences, U.S.A.)

differentially activated the right parietal region of the cerebral cortex, whereas those for Best Friend and Celebrity tended to activate instead the left lateral temporal region (Fig. 5).

These very interesting findings still leave open several theoretical explanations of how the self is instantiated in the brain, including, following Tulving's distinctions [47], the role played by episodic memories (e.g., "this happened to me on my birthday") versus that of semantic memories (e.g. "I was born in New York"). Episodic and semantic memories combine to form the category of declarative memories, which are contrasted with procedural memories, e.g. remembering how to perform a particular task. Now, Lou and his colleagues believe that their experiment sheds light on the nature of the self, in that episodic memories form the basis of what it "feels like" to be a particular person (since presumably those episodes in which we acted as agents combine to form the history of "our" experience). The instantiation of the self, thus, would likely be found in those structures that deal with episodic memories. Saxe et al. nearly concur, pointing to the fact that "theory of mind, self-reflection, and autobiographical episodic memories are correlated in child development," as determined by Moore and Lemmon [48].

Unfortunately, Saxe et al. do not realize that this PET study fails to reject the MPFC null hypothesis: that there is no significant difference in activation between the self-attribution and the other conditions. Fortunately for the MPFC hypothesis, a different experimental test may still provide support, for a Darwinian account casts doubt on the PET study's approach. To see why, we should consider briefly the theoretical context, from psychology, that gives plausibility to the interpretation of the Lou et al. experiment. Much of the work on self-knowledge of personality traits has been influenced by the demonstration by Rogers, Kuiper and Kirker [49] that self-referential questions (e.g. "Does the word *kind* describe you?") lead to better recall than questions involving more abstract judgments (e.g. "What does *kind* mean?"). Moreover, the distinction between procedural and declarative memory, according to Kline [50] coincides with Ryle's [51] distinction between *knowing how* and *knowing that.* Since philosophers tend to associate rationality with knowing that, and they have influenced psychologists, when it comes to the self of a rational creature, it is not surprising that we should try to account for the self in terms of declarative memories. Lou et al. narrow it further to episodic memories, appealing indirectly to the philosopher Thomas Nagel's famous "What is it like to be a Bat?" [52]

Considering knowledge in an evolutionary context, though, and particularly what sense of "know" should sensible apply to a brain in that context, Ryle's distinction is placed much in doubt, for knowledge is ultimately about action [20]. The neural perspective would also seem to apply to self-referential questions. In Plato's dialogues, Socrates goes around asking people to tell him what terms like "justice" or "knowledge" mean. People give him examples, many examples of how they use those terms, and he complains that he is asking for one thing and they are giving him many. Nevertheless, the brain works better in terms of examples that are related by "family resemblances" instead of abstract definitions [16]. Moreover, neural nets are successfully trained by examples to perform tasks such as indentifying faces from photographs, according to P.M. Churchland [53], but typically have no representations of any rules, nor do they "achieve their function-computing abilities by following any rules. They simply embody the desired function, as opposed to calculating it," p. 12. This is not to question the result that self-referential questions lead to better recall (presumably of episodic memories), but rather that since it should be so obviously expected, given that concrete examples are a lot easier to handle than abstract definitions, the significance of the contrast in recall found by Rogers et al. is not as great as some make it out to be.

A Darwinian account would lead us to suppose instead that the self should be instantiated in something akin to procedural memories, to preparedness for interaction with the world. To support this approach I proceed to criticize the interpretation of the Lou experiment on two fronts, theoretical and experimental (although in neuroscience these two categories are not mutually exclusive). I will first consider Stanley Klein's social neuropsychological approach to understanding the self [50]. Klein concentrates on one aspect of self-knowledge: knowledge of one's own personality traits, as Lou et al. do. As Klein explains, case studies show such knowledge is resilient even after damage to the ability to retrieve episodic memories and several kinds of semantic memory. For example, patients K.C. and R.J. had reliable knowledge of their personality traits (their self-appraisals were strongly correlated with those given by people who knew them well). K.C. lost all his episodic memories in a motorcycle accident and underwent a significant personality change afterwards. But even though he was completely unable to retrieve episodic memories, he gave very accurate descriptions of his post-morbid personality [54].

Interestingly enough, and of relevance to the emphasis on the subconscious defended in this paper, key self-knowledge – K.C.'s personality traits – was acquired and retrieved without *conscious* access to episodic memories. Klein considers that knowledge of personality traits comes from trait summaries and is a form of semantic memory. But those trait summaries, in the absence of episodic memories, would have to come by unconsciously (or subconsciously), for what would a new experience be *consciously* compared to in order to update his self evaluation? R.J.'s inability to retrieve episodic memories was developmental, that is, he probably never had any. This drastic reduction of the role of consciousness in the one characteristic of the self that would seem most likely involved with consciousness is very telling.

In any event, it is another patient who lost his hippocampuses to an operation, H.M., who provides a hint as to how memories, if they are going to become integrated into the self, must resemble procedural memories instead, i.e. they must prepare us to interact with our environment, which is the key evolutionary justification for having a self, as we have seen. H.M. could form no episodic memories at all. If you were introduced to him but left the room for a few minutes, upon your return you had to be reintroduced. One day his doctors performed an informal experiment on him. They had an orderly be rude to him. In future encounters, even though H.M. thought he was meeting that orderly for the first time, he showed apprehension towards him. He had learned to take a certain posture towards the man, just as was also able to learn some practical skills, without conscious memory of such training.

During our experimental critique, my colleagues Shunshan Li and Matthew Cole and I first tried to replicate in an fMRI scanner Lou's PET study.[3] Lou and his colleagues had a two-step procedure in order to create episodic memories. The subject would first sit at a computer console and answer questions about his or her personality traits, some likeable, some not (questions of the following sort: "Are you honest?" "Are you manipulative?"). A few minutes later the subject would go into the scanner and there would be asked to remember what he or she had answered at the computer console. These were presumably episodic memories because answering the questions in the computer were episodes in which the subject had performed as agent. We were unable to replicate the original study, but that might have been due to lack of ingenuity on our part, since transferring experiments across experimental modalities can be quite a challenge. Nevertheless our impression was that the very attempt to create episodic memories caused several confounds in our collection of data. The most serious one came up during our attempts to fine

[3]We performed our experiments on a 3 Tesla scanner in Mark Haacke's MR Research Laboratory at Harper Hospital of Wayne State University, in Detroit, Michigan. We were assisted by our students Christina Minta, Timothy Bond and Casandra Langley.

tune our experimental paradigm. All five subjects who participated in this phase (members of the team) reported the same problem: we really doubted that we were using our memories at all in most questions concerning our own personality traits. If you are like most people, when asked "Did you say you are honest?" You immediately say "yes" without truly "remembering" what you said simply because you automatically attach honesty to yourself. If at the computer console you were asked, "how much is 2 + 2?" You would answer "4." If later you are asked, "How much did you say 2 + 2 is?" Would you really exercise your memory at all to come up with the answer, or would you rather say "4" automatically because that sort of thing is already at your fingertips?

Given this experience in the laboratory, and being guided by the Darwinian considerations adduced above, we decided to eliminate the memory task altogether. A group of 15 volunteers, 8 men and 7 women, were simply asked questions about their personality traits, as well as the traits exhibited by their best friend and a celebrity (we chose Bill Gates). Our new hypothesis was that the brain would handle questions about self differently from questions about the two other conditions, and that the handling, at least in the cortex, would activate areas concerned with preparation for action. As of this writing, our *very preliminary* group analysis (Fig. 6) seems to support our hypothesis: the most significant area of activation of the Self condition is the right supplementary motor area, which is involved in the planning of actions, particularly in the planning of actions under internal control (as opposed to, say, an automatic response to a stimulus). If upon final review this result holds in contrast with Best Friend or Bill Gates we would have sufficient warrant to embark with other colleagues on a series of experiments, guided by similar Darwinian considerations, to determine the neural structures underlying the self.[4] At the very least, though, by directly comparing the attributions of traits to self and others, this experiment will offer a test of the hypothesis that so strongly connects the MPFC and the self.

Notice, incidentally, that the planning for action need not be conscious. Recent experiments, ably summarized by Custers and Aarts, show that subliminal clues, of which we are not conscious, can lead us to consider certain goals, to work out the possible means to attain the relevant outcomes, and to assess the value of the outcomes, i.e. whether they are rewarding. These results show that there are unconscious mechanisms for pursuing goals, since we can "unconsciously detect the reward value of a primed goal and prepare feasible actions that make that goal attainable" [55]. This is possible, among other things, because actions and their outcomes are associated, through prior learning, on a sensory and motor level.

[4]Our intent at this time is to do some more sophisticated versions of the present experiment with patients of mental illnesses that seem to affect the self – schizophrenia, for example, for the patients often believe that they are under someone else's control. I have also designed experiments to help determine the role of consciousness in volitional processes. And with Matthew Cole and our students we are developing a poker game to be played in a fMRI scanner to study gambling addiction.

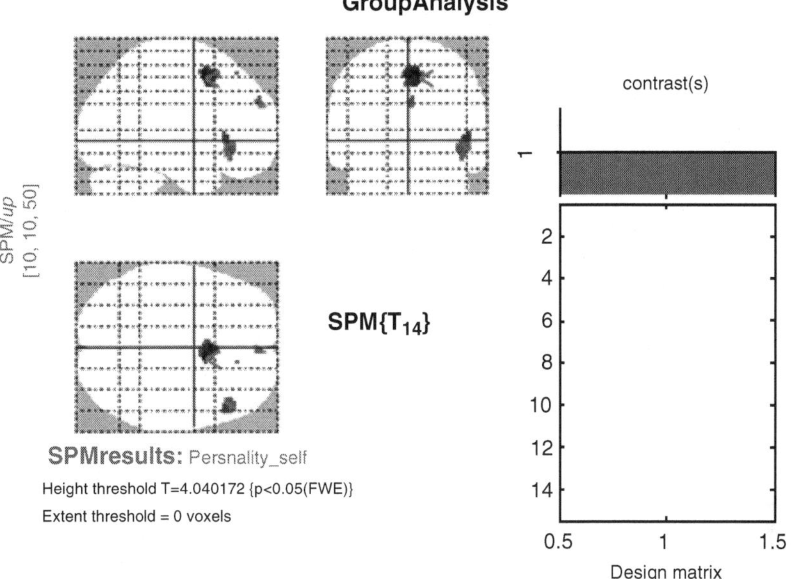

Statistics: *p-values adjusted for search volume*

set-level		Cluster- level			voxel- level					mm mm mm		
ρ	c	ρ carrected	k_E	ρ uncarrected	ρ PM E-corr	ρ FDE-corr	T	(z≡)	ρ uncarrected			
0. 0004		0.000	176	0.000	0. 000	0. 000	10.35	5. 42	0. 000	10	10	50
					0. 000	0. 000	9. 77	5. 29	0. 000	2	14	52
					0. 000	0. 000	9. 32	5. 18	0. 000	6	14	44
		0.000	91	0.003	0. 000	0. 000	8.85	5. 06	0. 000	52	26	4
					0. 000	0. 000	7.85		0. 000	42	32	−6
					0. 001	0. 000	6.96	4. 50	0. 000	48	32	−12
		0.011	12	0.217	0. 000	0. 000	7.36	4. 64	0. 000	4	56	34
		0.032	2	0.630	0. 001	0. 000	6.42	4. 31	0. 000	12	38	54

table shows 3 local maxima more than 8.0mm apart

Height threshold: T= 4.04, p=0.001 (0.050){p<0.05(FWE)} Degrees of freedom =[1.0,14.0]
Extent threshold: k=0 vaxels, p=1,00(0.050)
Expected voxels per cluster,<k> =8.452
Expected number of cluster, <c> = 0.05
Expected false discovery rate, <= 1.00

FWHM=11.69.011.5 mm mm mm, 5.8 4.5 5.8{voxels};
Volume: 2248; 281 voxels; 0.5 resels
Vocel size: 2.0 2.0 2.0 mm mm mm;(resel = 151.60 voxels)

Fig. 6 Preliminary results of an fMRI study on self-knowledge by the author and his colleagues. The supplementary motor area was significantly activated (vs. controls) when subjects considered whether a series of personality traits could be attributed to them. Data analysis is not yet complete, as of this writing

Thus in a game we kick the football a certain way, intentionally, even though we have no time to think (consciously) about it, to score a goal for which we will take full credit; and after the game we drive home, making decision after decision unconsciously, while concentrating on what we are going to have for dinner. In these and many other ways, the self, sometimes consciously and sometimes unconsciously, exercises its will, as it should.

References

1. Wegner, D.M.: The Illusion of Conscious Will. MIT Press, Cambridge (2002)
2. Nielson, T.I.: Volition: a new experimental approach. Scand. J. Psychol. **4**, 215–230 (1963)
3. Walter, W.G.: Presentation to the Osler Society. Oxford University, Oxford (1963)
4. Libet, B.: Unconscious cerebral initiative and the role of conscious will in voluntary action. Behav. Brain Sci. **8**, 529–539 (1985)
5. Blackmore, S.: Consciousness: A Very Short Introduction. Oxford University Press, Oxford (2005)
6. Watson, G.: Free will. In: Sosa, E., Kim, J. (eds.) A Companion to Metaphysics. p. 178. Basil Blackwell, Oxford (1994)
7. Munevar, G.: A naturalistic account of free will. Dialogos **33**(72), 43–62 (1998); reprinted as Ch. 12 of [20]
8. Munevar, G.: El cerebro, el yo, y el libre albedrío. In Guerrero, G. (ed.) Entre Ciencia y Filosofía: Algunos Problemas Actuales (Programa Editorial Universidad del Valle, Cali, Col 2008, 291–308); reprinted as: Apéndice al capítulo 12. *La* Evolución y la Verdad Desnuda (Ediciones Uninorte. Barranquilla, Col 2008, 253–278)
9. Llinas, R.: I of the Vortex. MIT Press, Cambridge (2001)
10. Dennett, D.C.: Freedom Evolves. Viking, New York (2003)
11. Damasio, A.: The Feeling of What Happens. Harcourt Brace & Co, New York (1999)
12. Ramachandran, V.S.: A Brief Tour of Human Consciousness, pp. 7–9. Pi Press, New York (2005)
13. LeDoux, J.: Synaptic Self, p. 27. Penguin, New York (2003)
14. Monroe, R.: Schools of Psychoanalytic Thought. Holt, Rinehart and Winston, New York (1955)
15. Edelman, G., Tononi, G.: A Universe of Consciousness. Basic Books, New York (2000)
16. Munevar, G.: Conquering Feyerabend's conquest of abundance. Philos. Sci. **69**(3), 519–536 (2002)
17. Hoffman, D.D.: Visual Intelligence, pp. 2–9. W.W. Norton, New York (1998)
18. Johnston, V.S.: Why We Feel, pp. vii–viii. Perseus Books, Cambridge (1999)
19. Koch, C.: The Quest for Consciousness. Roberts & Co., Englewood (2004)
20. Munevar, G.: Radical Knowledge. Hackett, Indianapolis (1981)
21. Munevar, G.: Evolution and the Naked Truth. Ashgate, Aldershot (1998)
22. Munevar, G.: Perception and natural selection. In: Frendo, H. (ed.) The European Mind: Narrative and Identity, pp. 222–232. Malta University Press, Malta (2010)
23. Crick, F.: The Astonishing Hypothesis: The Scientific Search for the Soul, pp. 265–268. Scribner's Sons, New York (1994)
24. Churchland, P.S.: Neurophilosophy. MIT Press, Cambridge (1986)
25. Gazzaniga, M.S.: The Mind's Past. University of California Press, Berkeley (1998)
26. Gazzaniga, M.S., LeDoux, J.: The Integrated Mind. Plenum Press, New York (1978)
27. Churchland, P.M.: The Engine of Reason, the Seat of the Soul. MIT Press, Cambridge (1996)
28. Churchland, P.M.: Neurophilosophy at Work. Cambridge University Press, Cambridge (2007)
29. Chemero, A., Silberstein, M.: After the philosophy of mind: replacing scholasticism with science. Philos. Sci. **75**(1–27), 22 (2008)
30. Baler, R.D., Volkow, N.D.: Drug addiction: the neurobiology of disrupted self-control. Trends Mol. Med. **12**(12), 559–566 (2006)

31. Volkow, N.D., Li, T.: Drug addiction: the neurobiology of behavior gone awry. Nat. Rev. Neurosci. **5**, 963–970 (2004)
32. Keenan, J.P., Wheeler, M.A., Gallup, G.G., Pascual-Leone, A.: Self-recognition and the right prefrontal cortex. Trends Cogn. Sci. **4**, 338–344 (2000)
33. Keenan, J.P., Rubio, J., Racioppi, C., Johnson, A., Barnacz, A.: The right hemisphere and the dark side of consciousness. Cortex **41**, 695–704 (2005)
34. Gallup, G.G.: Chimpanzees: self-recognition. Science **167**, 86–87 (1970)
35. Gallup, G.G., Anderson, J.L., Shillito, D.P.: The mirror test. In: Bekoff, M., Allen, C., Burghardt, G.M. (eds.) The Cognitive Animal: Empirical and Theoretical Perspectives on Animal Cognition, pp. 325–333. University of Chicago Press, Chicago (2002)
36. Morin, A.: Self-awareness and the left hemisphere: the dark side of selectively reviewing the literature. Cortex **43**, 1068–1073 (2007)
37. Macrae, G.N., Heatherton, T.F., Kelley, W.M.: A self less ordinary: the medial prefrontal cortex and you. In: Gazzaniga, M.S. (ed.) The Cognitive Neurosciences III, pp. 1067–1075. MIT Press, Cambridge (2004)
38. Craik, F.I.M., Moroz, T.M., Moscovitch, M., Stuss, D.T., Winocur, G., Tulving, E., Kapur, S.: In search of the self: a positron emission tomography study. Psychol. Sci. **10**, 26–34 (1999)
39. Kelley, W.M., Macrae, C.N., Wyland, C.L., Caglar, S., Inati, S., Heatherton, T.F.: Finding the self: an event-related fMRI study. J. Cogn. Neurosci. **14**, 785–794 (2002)
40. Platek, S.M., Loughead, J.W., Gur, R.C., Busch, S., Ruparel, K., Phend, N., Panyavin, I.S., Langleben, D.D.: Neural substrates for functionally discriminating self-face from personally familiar faces. Hum. Brain Mapp. **27**(2), 91–98 (2006)
41. Platek, S.M., Wathne, K., Tierney, N.G., Thomsona, J.W.: Neural correlates of self-face recognition: an effect-location meta-analysis. Brain Res. **1232**, 173–184 (2008)
42. Gillihan, S.T., Farah, M.J.: Is self special? A critical review of evidence from experimental psychology and cognitive neuroscience. Psychol. Bull. **13**, 76–97 (2005)
43. Turk, D.J., Heatherton, T.F., Macrea, C.N., Kelley, W.M., Gazzaniga, M.S.: Out of contact, out of mind: the distributed nature of the self. Ann. NY Acad. Sci. **1001**, 1–14 (2003)
44. Moran, J.M., Macrae, C.N., Heatherton, T.F., Wyland, C.L., Kelley, W.M.: Neuroanatomical evidence for distinct cognitive and affective components of self. J. Cogn. Neurosci. **18**, 1586–1594 (2006)
45. Saxe, R., Moran, J.M., Scholz, J., Gabrieli, J.D.E.: Overlapping and non-overlapping brain regions for theory of mind and self reflection in individual subjects. Soc. Cogn. Affect. Neurosci **1**, 299–304 (2006)
46. Lou, H.C., Luber, B., Crupain, M., Keenan, J.P., Nowak, M., Kjaer, T.W., Sackeim, H.A., Lisanby, S.H.: Parietal cortex and representation of the mental self. PNAS **101**(17), 6827–6832 (2004)
47. Tulving, E.: What is episodic memory? Curr. Dir. Psychol. Sci. **2**, 67–70 (1993)
48. Moore, C., Lemmon, K.: Self in Time: Developmental Perspectives. Lawrence Erlbaum Associates, Inc., Mahwah (2001)
49. Rogers, T.B., Kuiper, N.A., Kirker, W.S.: Self reference and the encoding of personal information. J. Pers. Soc. Psychol. **35**, 677–688 (1977)
50. Klein, S.: The cognitive neuroscience of knowing one's self. In: Gazzaniga, M.S. (ed.) The Cognitive Neurosciences III, pp. 1077–1089. MIT Press, Cambridge (2004)
51. Ryle, G.: The Concept of Mind. Barnes and Noble, New York (1949)
52. Nagel, T.: What is it like to be a bat? Philos. Rev. **LXXXIII**(4), 435–450 (1974)
53. Churchland, P.M.: A deeper unity: some Feyerabendian themes in neurocomputational form. In: Munevar, G. (ed.) Beyond Reason: Essays on the Philosophy of Paul Feyerabend, pp. 1–23. Kluwer, Dordrecht (1991)
54. Tulving, E.: Self-knowledge of an amnesic individual is represented abstractly. In: Srull, T.K., Wyer, R.S. (eds.) Advances in Social Cognition, 5th edn, pp. 147–156. Erlbaum, Hillsdale (1993)
55. Custers, R., Aarts, H.: The unconscious will: how the pursue of goals operates outside of conscious awareness. Science **329**, 47–50 (2010)

The Problem of 'Darwinizing' Culture (or Memes as the New Phlogiston)

Timothy Taylor

Introduction

There have been many debates, over more than two millennia, about the relationship of humans to nature: philosophical, theological, anthropological, historical, biological, and so on. Whether we perceive ourselves as partially or wholly belonging to nature, or being above it or beside it, must depend to an extent on the way nature is conceived – for instance, whether it is seen as a divine creation or a wholly contingent and more or less random playing out of extant laws of matter and movement. Our attitude also depends on whether we believe we have a purpose in relation to it.

As humans, we know we can have our own purposes; this may lead us to recognise, correctly or not, purposiveness elsewhere. We may also investigate the forms of our purposes: how purposiveness has arisen. But just by hoping to answer that question we would adumbrate the distinction between humans and non-cultural life forms. Wittgenstein said that it is 'easy to imagine an animal angry, frightened, unhappy, happy, startled. But hopeful? ... A dog can expect its master, but can it expect its master will come the day after tomorrow?' [1, Nos 358, 360]. The inferred difference in forms of life between human and canine, which the philosopher pondered, is produced by the use of a cultural grammar by the former – an infinitely labile system of signs tacking between words and things, states and intentions ([2]; and see [3] for the disciplinary context).[1]

[1] I should clarify at the outset that this is not intended as an admission of dualism; as a materialist concerned with material causes, I am concerned with the distinctive patterns and – potentially – the *de facto* autonomous logics of the nested ontological levels of material existence. What is at issue here are the ways that cultural objects, whatever they consist in, may bring patterning into the world, and what sort(s) of patterning it is.

T. Taylor (✉)
Department of Archaeology, University of Bradford, BD7 1DP Bradford, UK
e-mail: timtaylor@gmail.com

M. Brinkworth and F. Weinert (eds.), *Evolution 2.0*, The Frontiers Collection,
DOI 10.1007/978-3-642-20496-8_6, © Springer-Verlag Berlin Heidelberg 2012

The question I want to pose, as a prehistorian and archaeologist, is whether such distinctively human experiences and capacities arise seamlessly from background biology. That is to say, is intentionality an emergent property of complex biological systems, or should we look to another form of patterning and recognize the appearance of a new form of life in culture-using humans?

Dawkins and others who support his meme idea believe the former: where the influence of the replicators of organic nature (genes) fades, units of culture (memes) take over, ostensibly to continue the pattern of Darwinian competition. That the first are conceptually directionless (albeit anthropomorphized as 'selfish') while the latter are clearly directed (if not always under direct control) does not appear to present any special problems for supporters of the idea. The territory claimed as won seems immense, as it allows human nature to be tidied up without the need for an additional explanatory paradigm. I am unconvinced, and believe that we have to continue the challenging task of providing an explanation for, and understanding of, diagnostically human behaviour in specifically human terms. I am not rejecting biology, any more than a biologist rejects physics; rather I am making a case for a hierarchy of systems of patterning involving three, not two systems.

Humans are, of course, animals. But then we are also a concentration of atoms which, when separated and dissociated, are not living and which do not compete to live. The point is that some things made with, for example, carbon are not alive (carbon dioxide), other things made with carbon are (a shark). Sharks show variance over time of a sort that is not just a frequency issue, as it is with carbon dioxide, but a formal one. They evolve on the basis of genetic recombination. Humans, like sharks, are carbon based and evolving too. This makes us animals like them. But just as a shark is not just atoms so humans are not just genes. Nor does proposing memes as proxy extensions of genes make it any more plausible that we should view ourselves as merely animals.

In trying to make a case for a realm of artifice with its own distinctive, largely non-Darwinian generative processes, my aim is not merely to point out again what could be foundationally different about humans, and help emancipate some areas of culture theory from an unproductive and misapprehended biological reductivism (especially meme theory). I also want to defend Darwin's mechanism of descent with modification from teleological readings. Separating biology from culture, especially in terms of the identification of the relevant generative mechanisms for the (different) kinds of formal variance produced in each case may make this job easier. There are certainly similarities between human biological and cultural evolution, but neither the drivers nor the logic of development need be shared.

The Genealogy of the Meme

An apparently minor problem presented itself to Richard Dawkins in formulating his influential 1976 neo-Darwinist manifesto, *The Selfish Gene* [6]. This was that a technological trick as simple as a section of sheep bladder deployed as a condom could subvert the reproductive logic of heterosexual intromissive sex among humans. The conjunction of a technical element of human sexual culture with a biologically-evolved pleasure system meant that the core mission of 'selfish genes' contained, in our species at least, a contradiction. In short, genes for intelligence and invention could lead to behaviours that might result in innovators failing to pass on their DNA to a subsequent generation (that is, succeeding in a planned childlessness). However, also consciously, they could pass on their technological know-how, along, even, with ideological justifications for not producing progeny. Dawkins ended the meme chapter in *The Selfish Gene* by saying 'We are built as gene machines and cultured as meme machines, but we have the power to turn against our creators. We, alone on earth, can rebel against the tyranny of the selfish replicators' and 'we do so in a small way every time we use contraception' ([6], p. 201 and footnote to the second edition, p. 322). The qualifier 'small' here might easily deflect our attention away from the actual scale of the problem, as the formulation of a cultural unit idea was critical to the project of retaining human behaviour within nature. Dawkins put it like this:

> I think Darwinism is too big a theory to be confined to the narrow context of the gene ... I think that a new kind of replicator has recently emerged on this very planet ... It is still in its infancy, still drifting clumsily about in its primeval soup, but already it is achieving evolutionary change at a rate that leaves the old gene panting far behind. The new soup is the soup of human culture ... We need a name for the new replicator, a noun that conveys the idea of a unit of cultural transmission ... *meme* ... Examples of memes are tunes, ideas, catch-phrases, clothes fashions, ways of making pots or building arches. Just as genes propagate themselves in the gene pool by leaping from body to body ... So memes propagate themselves in the meme pool by leaping from brain to brain in a process which, in the broad sense, can be called imitation [6, p. 191f]

Using a machine analogy similar to those favoured by Paley, Dawkins added 'the computers in which memes live are human brains' [6, p. 197]. One neologism encourages another, and soon there was 'memeplex': if genes clump up in chromosomes, Susan Blackmore reasoned, then memes can be thought of as clumping together in 'self replicating meme groups' [7, p. 19f].

Dawkins was by no means the first person to try to conceptualize cultures in terms of units, whether isolated or clumped. Indeed, such a tendency was apparent in (particularly German) ethnology in the later nineteenth century, when cultural trait lists first began to be drawn up to characterize different peoples as if they were natural species. The idea that there were fixed and culturally diagnostic artefact types had been elaborated by Gustaf Kossinna and introduced to English-speaking archaeology by V. Gordon Childe:

> We find certain types of remains – pots, implements, ornaments, burial rites and house forms – constantly recurring together. Such a complex of associated traits we shall term a 'cultural group' or just a 'culture'. We assume that such a complex is the material expression of what today would be called a 'people' [8, pp. v–vi].

Although Childe progressively lost faith in this appealing simplicity, it did not deter others, such as the neoevolutionary anthropologist Leslie White, from developing the terminology along the same lines. White, adducing cultural phenomena every bit as heterogeneous as those on Dawkins's list, argued for a

> class of things and events dependent upon symboling: a spoken word, a stone axe, a fetich, avoiding one's mother-in-law, loathing milk, saying a prayer, sprinkling holy water, a pottery bowl, casting a vote, remembering the Sabbath to keep it holy ... These things and events constitute a distinct class of phenomena in the realm of nature. Since they have had heretofore no name we have ventured to give them one: *symbolates*. We fully appreciate the hazards of coining terms, but this all-important class of phenomena needs a name to distinguish it from other classes [9, p. 230f].

White's aim was to build a new and objective science of culture (and those he influenced, such as Lewis Binford, were later to make great play of adapting avowedly and explicitly 'scientific', Hempelian, hypothetico-deductive-nomological methods to archaeology with consequent rewards from the US National Science Foundation in terms of grant support for major projects):

> I smoke a cigarette, cast a vote, decorate a pottery bowl, avoid my mother-in-law, say a prayer, or chip an arrowhead. Each of these acts is dependent on the process of symboling; each therefore is a symbolate ... we may treat symbolates in terms of their relationship to one another, quite apart from their relationship to the human organism ... If we treat them in terms of their relationship to the human organism, i.e., in an organismic, or somatic context, these things and events become *human behavior* and we are doing *psychology*. If, however, we treat them in terms of their relationship to one another, quite apart from their relationship to human organisms, i.e., in an extrasomatic, or extraorganismic, context, the things and events become *culture* – cultural elements or cultural traits – and we are doing *culturology* [9, p. 233]

Contemporary with White, F.T. Cloak made the distinction between the *instructions* for action people had in their heads and the physical effects of these when operationalized in the *material* world:

> an i-culture builds and operates m-culture features whose *ultimate function* is to provide for the maintenance and propagation of the i-culture in a certain environment. And the m-culture features, in turn, environmentally affect the composition of the i-culture so as to maintain or increase their own capabilities for performing that function. As a result, each m-culture feature is shaped for its *particular* functions in that environment [10, p. 170].

It was typical for the American cultural ecology of this time to view the environment as an objective given to which human cultures adapted; simplistic, too, was Cloak's idea that instructions were copied, in a computational or information-theory manner.

This simplification, along with the proposition concerning two, reciprocating forms of culture (internal coding producing external practice), proved irresistible to Dawkins when he began revision of his meme formulation:

> I was insufficiently clear about the distinction between the meme itself, as replicator, and its 'phenotypic effects' or 'meme products' on the other. A meme should be regarded as a unit of information residing in a brain (Cloak's 'i-culture'). It has a definite structure, realized in whatever medium the brain uses for storing information. If the brain stores information as a pattern of synaptic connections, a meme should in principle be visible under a microscope as a definite pattern of synaptic structure. If the brain stores information in "distributed" form [...] the meme would not be localizable on a microscope slide, but still I would want

> to regard it as physically residing in the brain. This is to distinguish it from phenotypic effects, which are its consequences in the outside world (Cloak's 'm-culture') [11, p. 109].

In this revised formulation, Dawkins tried to nuance the problem of the identity of the meme unit: 'Memes may partially blend with each other in a way that genes do not. New "mutations" may be "directed" rather than random with respect to evolutionary trends' [11, p. 112]. There is an implication here that some, even many, memes somehow *might not* be directed: yet if we accept the terminology, then whatever memes are they surely arise, at least in part, from human intentions, and are intimately connected to human agency. This idea is peculiar but provides a rhetorical device for levering in the idea that memes respect evolutionary trends, and the evolution implied is presumed – in the absence of qualification – to be Darwinian. Dawkins elaborates as follows:

> The equivalent of Weismannism is less rigid for memes than for genes; there may be Lamarckian causal arrows leading from phenotype to replicator, as well as the other way round. These differences may prove sufficient to render the analogy with genetic selection worthless, or positively misleading [11, p. 112].

This reference to Lamarck and the doctrine of the Weismann barrier, which postulates that genes condition somatic development and never *vice versa* (itself coming under renewed critical scrutiny in the face of advances in epigenetic research), is at best disingenuous. As actual material objects typically serve as the primary model for subsequent production, to be observed in detail by craftspeople during the process of making replicas, modifications, improvements and innovations, the idea that causal arrows only *may* lead from phenotype to replicator rather misrepresents the case. Clearly they have to, whatever terminology we adopt.

In his objection to memetics, the anthropologist Dan Sperber focuses on the way memetics ducks the issue of intentionality, and picks up on the weakness in the idea of passively copying instructions: 'instructions cannot be imitated, since only what can be perceived can be imitated. When they are given implicitly, instructions must be inferred. When they are given verbally, instructions must be comprehended, a process that involves a mix of decoding and inference' [12, p. 171]; in similar vein, Anthony O'Hear suggests that the denial, implicit in the meme account, of the central human experience of mental reflection, is categorically fatal to that account [13].

Biology and Technology

When Tooby and Cosmides wrote that 'Human minds, human behavior, human artefacts, and human culture are all biological phenomena' [14, p. 21] they were making a claim (that human culture could be understood within an overarching, essentially neo-Darwinian, biological paradigm) with a rhetorical flair of which they must have been aware. Yet had an art historian made an equally inclusive reverse claim about the patterns of nature, they would not have been taken anywhere near as seriously. The anthropologist Maurice Bloch indulges in a similar thought experiment, imagining a sociologist, ignorant of Darwin and Mendel, inventing a novel term for the units of transmission of somatic characteristics in animals and attempting to foist it on biologists [15, p. 191]. This is really so little different from Dawkins trying to foist memes onto scholars across a range of disciplines without having – apparently – absorbed Peirce's contribution to semiotics or Wollheim's to aesthetic classification, that one might well pause to wonder how memes were ever taken seriously at all.

What has perhaps given the biological reductivists a degree of credibility is, firstly, the imprimatur of Charles Darwin (the brand, not the man) and, secondly, a sort of academic generosity at a point when Darwinism is increasing caught up in a face-off with crude religious fundamentalism. Clearly theologians who actually did (and do) deny evolution and make a claim of *ab initio* grand design on behalf of a deity were (and are) not taken seriously by many scientists. Humans are an animal species that has evolved and continues to evolve. Why should our behaviours be understood outwith the patterns of variance of other species, especially as the routines of learned behaviour among the more complex of them are increasingly well-documented and may appear to close the distance between us, at least along some parameters?

The anthropologist Ernest Gellner reasonably argued that the 'culture-proneness' of humans was (i) genetically enabled and (ii) comprised a programmed incompleteness or incapacity at birth [2]. That humans require years of learning to accomplish characteristically human tasks (such as basket making), whereas a spider has an instinct to complete characteristically arachnid ones (such as web-making), had been seen as a mark of natural incompleteness since at least the time of Herder. It was Herder too who encouraged us to imagine the distinctive and exclusive perceptual realms (*Umwelten*) of disparate species of animal and to consider whether there might be a similar kind of exclusivity in the most significant outer perceptions and inner experiences of different human communities – faiths, tribes, and social strata.

Before Herder, we thought of ourselves as aspiring to become cultivated, and saw culture as a unitary edifice. His innovation, introduced with the plural noun form, *Kulturen*, was to revive the idea, present in an acute early form in the ethnographic writings of Herodotus, of different peoples constructing different worlds for themselves [3, p. 31ff]. The nineteenth century saw focus shift away from the process by which any human might acquire culture as a universal quality towards a recognition that people are always encultured into specific cultures.

Much later, in the aftermath of the genetic synthesis of Darwinian theory, Gellner was able to note that what genetics underpinned was, when viewed across cultures, the greatest range of behavioural variance of any species on the planet; nevertheless, within any culture, behavioural deviations met with sanctions, including death. Globally, the species looked remarkable flexible, yet locally it was constructed of almost fascistically rigid units, each holding to a belief that its way of life was guaranteed by some form of natural right. A common judgment on enemies is that what they do is wrong because it is somehow a transgression against nature, not that it is different because they have encultured themselves with values antithetical to, and behaviours different from, our own, embedded in a coded material realm of clothes, and houses, tools and motifs so apparently coherent and all-immersive as to threaten belief in the continuity of our own constructs.

We do not just have the ability to form cultures; we are driven to do it. But if culture-proneness is clearly genetic, it is genetically enabled in a novel way, through underdetermination, facilitating an extreme version of the Baldwin effect.[2] Thus the individual cultures that come into being are not specifically genetically predicated to any great extent. This is clear enough simply from the historical and archaeological records, packed as they are with instances of revolutionary change in material life and sharp shifts in concomitant behaviours happening within genetically ongoing populations. Cultural 'phenotypes' are, at least proximally, the product of technology, language and art; whether they are ultimately genetically determined by an evolutionary, Darwinian logic is a moot point. Indeed, it might be that it is technological innovation that now significantly leads biology, with human physiological adaptation constantly playing catch-up.

In *The Artificial Ape* I argued for the essential veracity of the current archaeological and palaeontological chronology of evidence for tools use and appearance of genus *Homo* – that is, in that order [16]. Use of unmodified tools now provably extends back to 3.3 million years ago among small brained upright walking australopithecines; deliberate edge production (chipped stone tools that are plausibly for more precise and potentially more complex tool-related tasks) appears by 2.6 million years ago; *Homo*, however, does not provably appear before around 2.2 million years ago. Viewed in terms of muscular strength alone, we are the weakest of the great apes. Compensations are found in blades, slings, levers, spears and fire; both logic and the currently available empirical evidence suggest that many of these things preceded the sharp reduction in canines, the increase in cranial size, and the diminution of our innate capacities at birth (paedomorphism) that characterized the emergence of our genus. Our somatic deficits and our unprecedented intellectuality emerged together in the rain-shadow of technology. One conclusion of arguments that cannot be rehearsed here is that we did not start making things because we were growing ever smarter (due to some Darwinian mechanism that made larger brained mates more attractive); rather, the affordances of novel artefacts took the pressure off natural selection.[3]

The evolution of technology now outpaces natural selection so greatly that we have entered an era where biological variation is ever more frequently constrained by us; increasingly it will also be *designed* by our species. The technological realm reverberates in the biological one. But perhaps because the transfer of power from

[2]This is not the place to discuss the complexities of the potential relationship between, or partial identity of, behavioural plasticity in humans and the emergence (or definition) of free will.

[3]The question of canine reduction in hominins, known to long precede the appearance of chipped stone technologies, may well indicate an extensive phase of expedient use of found objects as tools, as emerging data from East Africa suggests.

the natural to the artificial is accelerating in both scale and range, we underestimate the fundamental shift that the first technologies brought, and even apparently simple technologies continued to bring, in certain parts of the world. That is, if my view challenges ideas about what happened in deep prehistory it may also be open to the objection that there are peoples alive today, or in recent history, who appear to have led a natural existence.

Darwin himself was of the opinion that the most basic technologies, especially those he saw among the Fuegians and the Australian aborigines, indicated a more or less structural connection to his placement of those peoples on a lower rung of his schematic ladder of human evolution from apes to humans. Meeting the inhabitants of Tierra del Fuego for the first time, he wrote in his *Beagle* journal: 'I could not have believed how wide was the difference between savage and civilised man: it is greater than between a wild and domesticated animal' ([17], entry for Dec. 17th 1832 = Chap. 10 and p. 197f in this edition). He continued the thought in *The Descent of Man*: 'the throughout the world'. The gradation he perceived, linking 'the negro or Australian civilised races of man will almost certainly exterminate, and replace, the savage races and the gorilla' with the 'baboon at the lower end' and the 'ever-improving' white or Caucasian race at the upper end would thereby be erased and, in evolutionary terms, 'the break between man and his nearest allies will then be wider' [18, p. 168f]. (Matt Ridley plausibly argues that it was the Fuegian experience much more than the Galapagos finches, that convinced Darwin of the reality of evolution [19].)

Evolution and Culture History

From at least the beginning of the nineteenth century, attempts have been made to typologize human societies, and to classify different forms of culture. This allowed a sequence of development or evolution for our own species to be envisioned in which the natural (somatic) and the artificial (extrasomatic) aspects were set out on the same field of play, in interplay and obedient to the same logic of development. Terminological traffic as well as the formulation of underlying concepts, especially in relation to generative aspects and ends (teleology) was two-way. Thus, when social Darwinism emerged in the wake of Darwin's own formulation of biological evolution, the potentially teleological phrase 'survival of the fittest' was reverse engineered into *The Origin of Species* through the persistence of Herbert Spencer. (Although, to be fair, it may have been latent in the 'favoured races' of Darwin's subtitle, a hangover from the intense intentionality he observed among his favoured pigeon fanciers.) In any case, the aimlessness (and perhaps even directionlessness) of descent with modification was compromised by the (re)introduction of a hint of fitness for purpose, an idea that Darwin had tried hard to turn his back on when he made the decisive break from Paley.

Whether or not Darwin meant to encourage a biologically innatist reading of human material culture, the result was that ethnographers subsequently elaborated a distinction between culturally creative peoples and *Naturvolk*, the latter little better than forest dwelling animals. Set at the very lowest end of the scale of human types were the Aboriginal Tasmanians, already essentially exterminated by the time the young Darwin reached recently-built Hobart. Unable to make fire, living naked and houseless in a cold and wet environment, their accomplishments (or lack of them) were long attributed to innate intellectual inferiority, consonant with a position lower down on the rising ladder of human evolution. Their simple stone tools were equated with those of the earliest hominin tool users of the Oldowan phase, or even with those of chimpanzees [16, pp. 33–54]. On a memic understanding, such as that of Laland and Odling-Smee, this would be an example of cultural stasis 'analogous to the elimination of genetic variation by stabilizing natural selection in population genetics' [20, p. 133].

Hitler subscribed to ideas of biological and social Darwinism through his belief in the Nordic 'Aryan' strain of humanity. The imagined Aryans were the postulated 'bearers' of the innovative Indo-European (or Indo-German) language and culture. By this point, then, Herder's *Kulturen* were thought of as the inheritors and transmitters of 'traits' every bit as distinctive as the short and long traits of Mendel's sweet peas. These traits were considered to be composed of pure units. The swastika (to pick a potent example) was one such unit. Claimed to be as recognizable on medieval Germanic pottery, Greek Iron Age pottery and in Indian temple friezes, the symbol was taken as a sign that the ancient Aryan master race had once occupied most of the Eurasian continent; Neolithic German invaders had brought culture (and the alphabet) to Greece from the north and gone on, in a prefigurement of Alexander, to carve an empire in the east that the Nazis would rebuild (Sanskrit, after all, was an Indo-European language, as was ancient Greek).

The swastika provides an interesting test of the meme concept; beyond the political dirty bath water there is the empirical fact that the phenomenon can be analysed into historically and cultural distinctive phenomena. That is, the form can be, and has been, arrived at in many ways. In India, a regular setting out of the nine principal deities as dots in a square allows the points to be dynamically integrated with two lines that dogleg to form the same shape that appeared as a variant of the 'Greek key' and probably arose from earlier curvilinear Bronze Age motifs; in Germanic Europe, the metal strap retainers on horses bridles were themselves made in the shape of horses' heads: four in a whirligig design around a central boss provide another route to the same familiar and infamous geometry. In reality, the sign is not one but many, found perhaps on every continent and in a wide range of cultures historically unconnected one to another. There is no definitive 'unit of culture' that is the swastika. The shape can be arrived at in a variety of ways. It can be named in different ways, and understood to mean different things. Particular forms can be reproduced or imitated but only through a process involving intentionality and the operation of cultural grammars – those rule-based generative systems typically carrying additional and/or modified content by virtue of developing context.

In Aunger's edited volume, *Darwinizing Culture*, Adam Kuper, critiquing the meme concept from an anthropological perspective ('yet to deliver a single original and plausible analysis of any cultural or social process' [21], p. 188), notes a claim made by Boyd and Richerson similar to that made by Tooby and Cosmides already cited: 'To the extent that the transmission of culture and the transmission of genes are similar processes, we can borrow the well-developed conceptual categories and formal machinery of Darwinian biology to analyse problems' [22, p. 31]. Kuper responds that 'it all sounds so pragmatic, so scientific, so reasonable that it is easy to forget that it is all a matter of metaphor and simile ... memes are rather shadowy entities' [21, p. 185]. He goes on:

> if memes are what we would normally call ideas (and, perhaps, techniques), then it is surely evident that ideas and techniques cannot be treated as isolated, independent traits ... [and are] are transmitted and transmuted in ways that are very different from the transmission of genes' [21, p. 187].

Another dissenting voice in Aunger's volume is Maurice Bloch, who makes the same point in relation to, among other things, catchy tunes (one of Dawkins's original examples): 'At first, some [memes] seem convincing as discrete units [however] on closer observation, even these more obvious 'units' lose their boundaries. Is it the whole tune or only part of it that is the meme?' [15, p. 194]. Behind these concerns is an important technical issue in philosophy, that of the distinction between tokens and types.

As Richard Wollheim long since pointed out, whistling a catchy tune that others recognize is a token of a type [23]. The reality is that that tune, in that context, with that performer and those hearers, behind whose recitation, or course, may lie conscious intentions and/or unconscious desires, specific to the time and place (or not), should be considered as a token: a specific concrete instance, standing in some relation to its type as a organism does to its species. But the parallel is inexact. The type might be thought to correspond to a gene and, in a cultural context to a 'meme'. Except that there is no original.

Wollheim demonstrates this vividly by asking about the whereabouts of the original of Beethoven's fifth symphony. The original might be claimed to be the first performance, or the conception of the symphony as Beethoven first imagined it, or as he completed the notation on paper, or a copy of the original (meaning first) publication of the score; it is also certain that 'it', were it decided upon, cannot be replayed: perceptions of the symphony that are as fresh and specifically informed as they were to the audience at the première are no longer possible. The modern musicologists perhaps best able to think themselves back into a state of pre-fifth symphony unknowing must also be those who are most acutely aware of its subsequent formative effect on the later development of classical music, and the haunting echoes in Brahms, Bruckner and Wagner.

As Beethoven's fifth symphony is a time-factored work of art, it is unclear whether there was ever a specific time when it all existed. One might assert that the opening (*Der der der dum!* [pause] *Der der der dum!*) is a sort of catchy fragment which we could call a meme; and Dawkins in fact asserts such a thing in relation to the basic 'Ode to Joy' theme in the ninth symphony: 'sufficiently distinctive and memorable to be abstracted from the whole symphony' as an aural logo for a German radio station [6, p. 195]. But what does it explain? Even these fragments of the tokens of types such as 'Beethoven's fifth' and 'Beethoven's ninth' can mean all sorts of things, in all sorts of contexts (with one deployed to critique biological reductivism from an archaeological perspective; the other to badge a radio station, or to support the idea of memes). Michael Barber, in outlining the views of the Viennese phenomenologist Alfred Schutz, notes that

> Music, differing from language in being non-representative, lends itself to phenomenological analysis in the meaning it carries beyond its mere physical nature as sound waves and in its character as an ideal object that must be constituted through its unfolding stages, i.e., polythetically [24].

Monothetic Versus Polythetic

Although the philosophy of categories is complex, most philosophers maintain that a sensible distinction can be made between monothetic and polythetic classification. The former designates a process of sorting into types where a particular attribute displayed by an individual is both sufficient and necessary for group inclusion. For example, diamond is a material that requires a particular arrangement of carbon atoms; if the carbon atoms are organized in this way, the material is classified as diamond; if they are not, it is not. The latter designate a process in which types are recognized in the absence of attributes that are at once sufficient and necessary for group inclusion. Cultural artefacts provide obvious examples (and it will be argued below that all cultural artefacts are polythetic entities).

It is true that, to a degree, there are also problems with the monothetic classification for biological taxonomy (the intellectual territory is complex and a few brief pointers to other work must suffice at this point). The Biblical injunction that each creature breed with its own kind (an essentialist model of a natural species) still colours conceptions even as the reality of descent with modification destabilizes the entities. Ernst Mayr [25] usefully showed how Buffon's nominal approach – admitting only individual existence as real and species names as heuristic only – gave way to the more nuanced biological species concepta we now use and argue over. Buffon wrote that 'the more we increase the divisions in the productions of nature, the closer we shall approach to the true, since nothing really exists in nature except individuals.' (English translation in [26], p. 160). While Charles Bonnet, developing his pre-Darwinian concept of evolution, wrote 'there are no leaps in nature. Everything in it is graduated, shaded. If there were empty space between any two beings, what reason would there be for proceeding from the one to the other?' (English translation in [26], p. 16) In this view, intermediate productions are always possible and, as Foucault notes, in this metaphysics 'only continuity can guarantee that nature repeats itself and that structure can, in consequence, become character [26, p. 160].

The philosophical difficulty of this – the antagonism between 'fixism' and 'evolutionism' that Foucault sees as irreducible yet complementary aspects of the classificatory schemas of the Enlightenment (a.k.a. the Classical age: [26], p. 164) – was at least recognized closer to the time by Kant, whose proposition that the touchstone of any species definition must be actual reproductive mate choices centres on the notion of conjunctions being foundational, shifting the emphasis away from static forms arrangeable in graded sequence back towards a classically Aristotelian, time-factored process. His *Realgattung* concept has similarities with the later *Formenkreis* idea of clinal variation in breeding populations. As this implies non-interbreeding populations, it hints at the relational definition at the core of Mayr's later formulation of distinct and bounded breeding pools.

But the direct influence was that of Illiger, just a generation before Darwin, whom Mayr cites *in extenso* (in translation). Of special importance here is Illiger's 1799 judgment that 'We can determine the species only on the basis of reproduction, and it is an error if one assumes, as is usually done, that the species originates through the extraction of common characters shared by several individuals. One has fallen into this error because one confused the species itself with the diagnostic characters of the species which the naturalist needs for his system, and because one thought that one had to apply the definitions of species and genus that are used in logic to organisms as well' [25, p. 169]. In the conception Illiger was criticizing (as Foucault neatly phrases it)

> the language of things would be constituted as scientific discourse by its own momentum. The identities of nature would be presented to the imagination as if spelled out letter by letter, and the spontaneous shift of words within their rhetorical space would reproduce, with perfect exactitude, the identity of beings with their increasing generality ... general grammar would ... be the universal *taxonomy* of beings [26, p. 161].

Mayr argued that Illiger's rejection of dichotomous Linnaean keys was part of his recognition that empirical reality could not be made to comply with traditional methods of logic; but it was more formatively Adanson in 1772 who paved a way for Beckner in 1959 to state that taxa were actually polytypic (as discussed by Needham [27]).

Sokal and Sneath developed the alternative term polythetic in 1963 and it was this that Clarke adopted for archaeology in 1968 as a label for a taxonomy of cultural entities in distinction from biological ones, viewed as monothetic [25, p. 169, fn 7; 28, 22ff; 29, p. 13ff; 30, p. 35ff]. This is potentially confusing. Some confusions are internal and latent in the texts; and it is also clear that the aims of these authors differ, along with their noncoincident subject matters. Before investigating that, it will be useful to provide the formal taxonomic definition of the difference between the two types of entity.

Adanson had suggested the idea of

> an arrangement of objects or facts grouped together according to certain given conventions or resemblances, which one expresses by a general notion applicable to all those objects, without, however, regarding that fundamental notion or principle as absolute or invariable, or so general that it cannot suffer any exception (English translation from preface to 1845 edition, as given in [26], p. 156).

This captured the essence of the monothetic–polythetic distinction. In David Clarke's words, and following Sokal and Sneath, a monothetic group is 'a group of entities so defined that the possession of a unique set of attributes is both sufficient and necessary for membership' [30, p. 35f]. This contrasts with a polythetic group which is constituted 'such that each entity possesses a large number of attributes of the group, each attribute is shared by large numbers of entities and no single attribute is both sufficient and necessary for group membership' [30, p. 36]. Sokal and Sneath were aware that 'it is possible that they are never fully polythetic because there may be some characters (or genes) which are identical in all members of a given taxon.' [29, p. 14].

Clarke's point was that biological entities – particularly animals species, for example – could better approximate monothetic entities: a backbone might be a sufficient and necessary attribute of a member of the vertebrata; lion DNA would do it for lions. The fact that, following Adanson, Illiger and Mayr, we know it to be more messy than this does not vitiate the sense of the general contrast with the artefactual products of technology. The critical issue for Clarke was that the monothetic method of typologizing by diagnostic key, whether or not it could still be made to work well enough for biologists in some or even most practical contexts, was inappropriate for artefacts. Human material culture items were, according to Clarke, clearly and fundamentally polythetic, yet they had been treated by scholars such as Kossinna, Childe and White as if they were biological species.

A chair, for example, might be thought about at first as if it is a natural type, like diamond. It should have four legs, a seat for sitting on, and a back part for leaning back against. But it should be immediately obvious that a chair can have three legs, or a single pillar leg, or rockers, or be fixed directly onto a raised surface without legs; conversely, a table may also have four legs, as may an electrical equipment stand. So, even if we were to decide that a chair must have four legs, this would not be an attribute *sufficient* in itself to sort an object into the chair category. But, as chairs patently do not require four (and just four) legs, so these are not a *necessary* attribute either; in fact, in some special contexts, no legs are necessary at all for something to be a chair. Drifting along various dimensions of variance, what we call chairs grade across towards other categories of seating: a bench seat, fold-down seating, aircraft ejector seats, and so on. It turns out that what we mean, in attribute terms, by a 'chair' in modern English-language speaking circles, represents an expectation of sorts. This can be met in a number of ways, but some are more expected and familiar than others. Thus many, in fact most, chairs have four legs. But the overall category, when represented in an attribute matrix, demonstrates a fuzzy edged intersect. As Needham neatly summarizes, viewed this way, while 'analysis is made more exact, comparison is made more intricate and difficult' [31, p. 60].

If the i-culture for 'chair' really was memically simple, then why would we have chair designers, commanding high fees for their original works? One could, of course, eschew them, as happened in the Soviet Union; but those with direct experience of what happened to chairs there, purified from a perceived bourgeoise, thus deviant, obsession with 'style' will know that the result was not only ugly, but frequently hopeless for sitting on. Nevertheless, it does go to show that the idea that there *ought* to be a chair meme as a sort of Platonic type has appeal not only to neodarwinists but to other ideologists who wilfully ignore the true level of complexity involved in artefact production.

The social anthropologist Tim Ingold notes that 'to resolve the paradox of distinction and continuity, we need to find a mode of human understanding that starts from the premiss of our engagement with the world, rather than our detachment from it' [32, p. 94]. Engagement occurs, for example, every time natural categories are translated into cultural ones. If I ask to be given a real diamond, I am asking for an object made of a material that can be monothetically defined; this can give the misleading impression that perhaps artefact categories can be monothetic after all. But it is not too difficult to see how any objection can be taken care of: clearly, I could be passed something comprising a minimum number of atoms in an arrangement that, chemically, would be diamond, but it would be such fine dust that I would not recognize it as 'a diamond'. There is an idea in my mind when I ask for a diamond, similar to that in my mind when I ask for a chair. Just as the chair has to be a solid material object (usually; but I might request a virtual chair in some contexts), so a diamond has to be made of diamond. But that is not precise enough for it to be necessarily 'a diamond'. It turns out that the most apparently monothetic aspect of cultural productions lies in naming and this depends, *pace* Wittgenstein, on agreement on expectations between at least two subjects involved in a language game. The tokens used in language games are context sensitive. Being given a diamond as a pre-marital pledge, and being given a diamond during a game of cards involve different expectations and materials, notwithstanding a complex skein of cultural genealogy linking the two symbols. Unlike a member of a biological taxonomy 'the' swastika is a polythetic entity. As is 'the' chair, 'the' car, 'the' basket, 'the' mug, 'the' house, and a diamond.

Even the simplest stone tools are only adequately defined polythetically. The Tasmanians, although recently reinstated as fully anatomically (and intellectually) modern *Homo sapiens*, have still struggled to be exonerated from a claim of some sort of maladaptive backwardness. Jared Diamond, in particular, shifted the explanation for their cultural deficits from biological to geographical determinism, arguing that long isolation caused them to lose useful adaptive traits, like knowing how to make fish hooks, or fire [33]. But there is an approach that shows Tasmanian material culture to be highly refined. Rather than not having 'progressed' since the Oldowan, Tasmanian stone tool production, outwardly similar to that of the Lower Palaeolithic period, can be seen to have involved a conscious rejection of multipart tools and curated (that is retained) objects. There is no space here to detail the rigorous logic of Tasmanian technological expedience, so different in sentiment and concept from the heavy entailment of the Europeans who were confronted with it at contact ([16], Chaps. 2, 6 and 8). Suffice to say, looked at in Cloak's terms, the *m-culture* looks the same, but the basis for the *i-culture* instruction is framed (*pace* Sperber [12]) by different goals and intentions; that is, it is not the same instruction.

Conclusion

My belief in technology as the key part of the environment, constraining how descent with modification occurs for humans does not mean that I exempt our species from the facts of biological evolution as it happens. It is not the mechanism of change but the generally accepted causes that I challenge. My position is, thus, almost the inverse of the Dawkins one when it comes to humans, yet consistent with evidence that Darwin himself would have taken seriously. As Ingold mordantly remarked 'Darwin was no Darwinist, let alone a neo-Darwinist, and he was a great deal more sensitive to the mutualism of organism and environment than many of those who nowadays yoke his name to their cause' [32, p. 97]. 'Environment' for humans rapidly became an interplay between three systems: inanimate nature, animate nature, and animated/animatable technology. I see technology as formative and able to undermine reproductive imperatives, and believe this is due to the emergence of a new form (or forms) of patterning. Dawkins sees technology as vital too (accepting that it can 'rebel' against genes, for instance) but he thinks this is a part of a universal Darwinism, essentially an extension of biological order and process. It is his contention that memes, as a cultural counterpart of genetic information, should have genealogies that conform to the Darwinian logic of descent with modification, selection, and survival of the fittest. The question, as set up here, becomes one of whether such logic, acceptable for inanimate natural entities of monothetic type, and roughly applicable in biological taxonomy, can also adequately classify variance among polythetic types. If it cannot, then on these grounds alone, the meme concept must be rejected.

I believe that three fundamental patterns of formal variance in the object world are known to us, viz: (i) inanimate systems, which involve the interplay of forces with natural physical hierarchies of monothetic (fundamental) entities; (ii) animate systems, which involve natural Darwinian competition between biological individuals belonging to species that approximate monothetic entities, and (iii) material culture systems, which involve the artificial generation of variance among polythetic entities. This idea is not wholly novel; Kevin Kelly, for example, talks of 'The Technium' [34]; that I speak of 'System 3' is not difference for the sake of it, but because Kelly understands technology to be the advanced manifestation of a deeper teleology, a force of creation that encourages complexity (or 'autocreation'), running seamlessly from inanimate nature through biological evolution and onwards [34, p. 355]; this I find much harder to accept.

We have already seen that the idea of neat, unitary memes is incompatible with the polythetics of formal variance among artefacts. But memes should also be rejected from a generative perspective. Going back to Kant's concept of *Realgattung,* it can hardly be maintained that artefacts mate with one another. How they do receive their forms is highly variable and complex, and there is unlikely to ever be one single law or overarching theory that explains all the behaviour and variance observed.

Nevertheless, particular phenomena are coming into better focus, including that of skeuomorphism, in which I take a special interest. This is the phenomenon whereby a vestigial feature is maintained through force of expectation when a change in construction material has to be effected for technological or cost reasons. A clear example would be a standard issue office desk which still looks as though it is made of wood. Even though we are not for a moment fooled by the laminate, we are comfortable with a properly met expectation. The interesting thing about this phenomenon is its unplanned outcomes, which I have described in more detail elsewhere in relation to the sequence that runs from upper Palaeolithic carvings of the human head in mammoth ivory, via the perfectly spherical elephant ivory billiard balls with which Darwin was familiar, to their synthetic material substitute as ivory became scarce and expensive, and the chemical innovation and formal production know-how that went on to have application both in synthetic ball bearings in the knee joints of Honda's walking 'Asimo' robots, and in the little plastic model balls that attached to struts and helped the discoverers of DNA to visualize its shape ([16], Chap. 7, and p. 202ff). The meme concept does not help the analysis in any way here. Precisely what is being copied? What do people think they are copying? And what, actually, are the entities brought into being? A memetic solution would immediately reveal to us the 'hazards of trying to theorise taxonomically about classes of facts that in empirical terms are polythetic' [27, p. 365].

I began this essay citing Wittgenstein whose concept – as Rodney Needham articulated – of 'family resemblances' was a polythetic concept, as critical for an analytical comparative anthropology as Clarke argued it to be for archaeology. It may also be that, *pace* Adanson and followers, it is critical for biology too. But 'replicants' are, definitionally, monothetic, otherwise they have no sense. Genes, in Dawkins's view, are the prime replicants. Their ability to preserve identity and vie with one another down the generations of somatic production is what underpins the only 'selfishness' they can ever have. The assertion that culture, too, breaks down into bounded units of competing imitation, espoused as a means to expand the Darwinian project, is not merely unhelpful in my field, but counteranalytic (and potentially counterproductive in terms of academic collaboration and constructive interdisciplinarity).[4]

Melville, prefiguring concerns that would become articulated, in distinctively different ways, by philosophers such as Husserl and Dennett, cultural ecologists such as Cloak, and Darwinists like Dawkins, wrote 'O Nature, and O soul of man! how far beyond all utterance are your linked analogies! not the smallest atom stirs or lives on matter, but has its cunning duplicate in mind' [36, p. 340]. Perhaps it is the technology of writing itself that embodies the cunning duplicitousness that only

[4]It may be worth reiterating at the close that this does not signal my dismissal of any and all Darwinian approaches in archaeology; many fruitful and fascinating avenues are opened by such research (see [35], for a useful overview); simply, memetics does not appear to me to be one of them.

language can create. Between the things in the outer world and words in the mind intentions arise whose generative power when acted upon, if not yet consistently greater than that of random mutation and descent with modification, contains a new level of complexity.

Acknowledgements I would like to thank my wife Sarah Wright for constructive criticism and masters students Emily Fioccoprile and Michael Copper; the shades of two former mentors – Ernest Gellner and Rodney Needham – are discernible.

References

1. Wittgenstein, L.: In: von Wright, G.H., Nyman, H. (eds.) Last Writings on the Philosophy of Psychology: Preliminary Studies for Part II of Philosophical Investigations, vol. I. Blackwell, Oxford (1982)
2. Gellner, E.: Culture, constraint and community: semantic and coercive compensations for the genetic under-determination of *Homo sapiens sapiens*. In: Mellars, P., Stringer, C. (eds.) The Human Revolution: Behavioural and Biological Perspectives on the Origins of Modern Humans, pp. 514–525. Edinburgh University Press, Edingburgh (1989)
3. Kuper, A.: Culture: The Anthropologists' Account. Harvard University Press, Cambridge (1999)
4. Aunger, R. (ed.): Darwinizing Culture: The Status of Memetics as a Science. Oxford University Press, Oxford (2000)
5. McGrath, A.: Dawkins' God: Genes, Memes and the Meaning of Life. Blackwell, Oxford (2005)
6. Dawkins, R.: The Selfish Gene, 2nd edn. Oxford University Press, Oxford (1989)
7. Blackmore, S.: The Meme Machine. Oxford University Press, Oxford (1999) (foreword R. Dawkins)
8. Childe, V.G.: The Danube in Prehistory. Clarendon, Oxford (1929)
9. White, L.A.: The concept of culture. Am. Anthropol. **61**(2), 227–251 (1959)
10. Cloak, F.T.: Is a cultural ethology possible? Hum. Ecol. **3**, 161–182 (1975)
11. Dawkins, R.: The Extended Phenotype: The Gene as the Unit of Selection. W.H. Freeman, Oxford (1981)
12. Sperber, D.: An objection to the memetic approach to culture. In: Aunger, R. (ed.) Darwinizing Culture: The Status of Memetics as a Science, pp. 163–173. Oxford University Press, Oxford (2000)
13. O'Hear, A.: Beyond Evolution: Human Nature and the Limits of Evolutionary Explanation. Oxford University Press, Oxford (1999)
14. Tooby, J., Cosmides, L.: The psychological foundations of culture. In: Barkow, J., Cosmides, L., Tooby, J. (eds.) The Adapted Mind: Evolutionary Psychology and the Generation of Culture, pp. 19–136. Oxford University Press, Oxford (1992)
15. Bloch, M.: A well-disposed social anthropologist's problems with memes. In: Aunger, R. (ed.) Darwinizing Culture: The Status of Memetics as a Science, pp. 190–203. Oxford University Press, Oxford (2000)
16. Taylor, T.: The Artificial Ape: How Technology Changed the Course of Human Evolution. Palgrave Macmillan, New York (2010)
17. Darwin, C.: The Voyage of the Beagle: Journal of Researches into the Natural History and Geology of the Countries Visited During the Voyage of HMS Beagle Round the World, under the Command of Captain FitzRoy, RN (1845). (Wordsworth Classics, London, 1997)
18. Darwin, C.: On the Origin of Species by Means of Natural Selection, or the Preservation of Favoured Races in the Struggle for Life. John Murray, London (1859)

19. Ridley, M.: The real origins of Darwin's theory, The Spectator, Wednesday 23 Sep (2009). http://www.spectator.co.uk/essays/all/5357791
20. Laland, K.N., Odling-Smee, J.: The evolution of the meme. In: Aunger, R. (ed.) Darwinizing Culture: The Status of Memetics as a Science, pp. 121–141. Oxford University Press, Oxford (2000)
21. Kuper, A.: If memes are the answer, what is the question? In: Aunger, R. (ed.) Darwinizing Culture: The Status of Memetics as a Science, pp. 175–188. Oxford University Press, Oxford (2000)
22. Boyd, R., Richerson, P.J.: Culture and the Evolutionary Process. University of Chicago Press, Chicago (1985)
23. Wollheim, R.: Art and Its Objects: With Six Supplementary Essays, 2nd edn. Cambridge University Press, Cambridge (1980)
24. Barber, M.: Alfred Schutz, Stanford Encyclopedia of Philosophy. Stanford University, Stanford. http://plato.stanford.edu/entries/schutz (2010)
25. Mayr, E.: Illiger and the biological species concept. J. Hist. Biol. **1**(2), 163–178 (1968)
26. Foucault, M.: The Order of Things. Routledge, London (2002)
27. Needham, R.: Polythetic classification: convergence and consequences. Man (NS) **10**, 349–369 (1975)
28. Beckner, M.: The Biological Way of Thought. Columbia University Press, New York (1959)
29. Sokal, R.R., Sneath, P.H.A.: Principles of Numerical Taxonomy. W.H. Freeman, San Francisco (1963)
30. Clarke, D.L.: Analytical Archaeology, 2nd edn. Methuen, London (1978)
31. Needham, R.: Remarks and Inventions: Skeptical Essays about Kinship. Tavistock, London (1974)
32. Ingold, T.: The evolution of society. In: Fabian, A.C. (ed.) Evolution: Society, Science and the Universe, pp. 78–99. Cambridge University Press, Cambridge (1998)
33. Diamond, J.: Guns, Germs, and Steel. Chatto & Windus, London (1997)
34. Kelly, K.: What Technology Wants. Viking, New York (2010)
35. Bentley, R.A., Lipo, C., Maschner, H.D.G., Marler, B.: Darwinian archaeologies. In: Bentley, A., Maschner, H.D.G., Chippindale, C. (eds.) Handbook of Archaeological Theories, pp. 109–132. Altamira Press, Lanham (2008)
36. Melville, H.: Moby-Dick or, The Whale. Penguin, London (1992)

Part II
Impact of Darwinism in the Social Sciences and Philosophy

Evolutionary Epistemology: Its Aspirations and Limits

Anthony O'Hear

Modern Epistemology

In modern epistemology the starting point is the isolated individual – in his or her stove-heated room or insulated study – alone with his ideas and experiences. The philosophical task is construed as one of moving from there to the external world and the things we normally take for granted, including induction. In their different ways, Descartes and Hume demonstrate the impossibility of success, given the starting point. From the Cartesian point of view, given the failure to extract a benevolent God from the Cogito, ideas alone cannot verify themselves, and so we are just left with ideas. Similarly Hume demonstrates by default that experiences prised off from the world are too slender a basis for what we want (a world of regularity existing apart from us).

It could (and should) be argued that the Descartes-Hume starting point misconstrues thought and experience. The private language argument suggests that there are necessarily public aspects to thought (or at least to thought which is linguistically dependent). Phenomenology suggests that experience is not as empiricism conceives it. We are not passive receivers of impressions or sense data, but are from the start actively engaged in a public world. We are, from the start, in the world, as agents, and not the externally related, would-be knowers of classical epistemology. The epistemological gap between self and world is a myth, though once it has been hypostasised it is unbridgeable, and scepticism becomes inevitable.

A. O'Hear (✉)
Department of Education, University of Buckingham, Buckingham MK18 1EG, UK
e-mail: anthony.ohear@buckingham.ac.uk

M. Brinkworth and F. Weinert (eds.), *Evolution 2.0*, The Frontiers Collection,
DOI 10.1007/978-3-642-20496-8_7,

Evolutionary Epistemology

Evolutionary epistemology (EE) suggests another starting point, which dissolves empiricist scepticism by not allowing it to begin. In essence, we reliably know the world (up to a point) because we have been moulded by the world to survive and reproduce in it. We are necessarily and immediately acting in the world (to survive and reproduce). We are not the passive and disconnected knowers of classical epistemology. In the struggle for survival, creatures with sense organs and conceptual schemes too unreliable have a praiseworthy tendency to die out, to be supplanted by those of their fellows with better mechanisms.

The mere fact we (and other creatures) have survived and reproduced for a while is some vindication of our (their) ideas about the world (or, if you like, shows that these have survived severe testing). EE will not produce a head-on refutation of scepticism, nor does it solve the problem of induction at a stroke. It shows that, given our survival, there must be something right about what we believe and act on, but only until the next challenge, when it might be shown even fatally, that our past solutions are no good for the new situation. This is the common logic of evolutionary explanation: a retrospective analysis of why past solutions worked to the extent they did, but always in a comparative sense. They did not have to be perfect but only good enough and, in particular, better for the purpose than those of competitors. This could suggest a fruitful line of investigation of our sensory and intellectual apparatus, showing how they latch on to useful features of the environment and exploit certain coincidences between our organs' receptivity and wavelengths and other features of our environment. But the evolutionary perspective is also one in which survivors only have to do better than actual competitors. So given the absence of any very well honed competitors, evolutionary success is compatible with quite a high degree of lack of perfect fit in the engineering sense.

Choice of Starting Point: Methodological Considerations

It could be argued that EE is not really epistemology at all, but really an avoidance of the hard epistemological problems. Specifically, in doing EE we are assuming that the theory of evolution is true, which in turn assumes that very many of our beliefs about the external world are true at all sorts of levels, so scepticism doesn't get tackled at all. This argument has some bite, *if* we are operating within a Cartesian or empiricist perspective, but much less so if we are not. We could ask, against the sceptic, why should we adopt this perspective, especially given the criticisms of it suggested above? It is a question of our philosophical starting point, around which a whiff of arbitrariness will always obtain once we see that what we take for granted does not have to be so taken. It is difficult to see knock-down arguments or conclusive proofs being available at so fundamental a level. So what we should consider at this point is the explanatory power and fruitfulness of competing starting points. It is certainly a possibility that the naturalistic presuppositions of EE will turn out to be a more fruitful starting point for an investigation of our knowledge than the isolated individual mind of classical epistemology.

Evolutionary Epistemology: Limitations

'What a biologist familiar with the facts of evolution would regard as the obvious answer to Kant's question was, at that time, beyond the scope of the greatest of thinkers. The simple answer is that the system of sense organs and nerves that enables living things to survive and orient themselves in the outer world has evolved phylogenetically through confrontation with and adaptation to that form of reality which we experience as phenomenal space' [1]. We should perhaps add to what Lorenz says 'and reproduce'. So, in developing what has some claim to be thought of as a genuine Copernican revolution in epistemology – decentering the human perceiver – the story EE tells is that our sense organs and cognitive faculties have been adapted to the phenomenal world as a means of our surviving and reproducing.

While the change of standpoint may indeed be productive and overdue, the first problem this account presents is the purely logical point that beliefs can be useful (to survival and reproduction) without actually being true. The aims are different, and even if epistemology isn't going to be foundational, we still want it to tell us about the truth of our beliefs, to explain why they are true, perhaps. EE may correctly say that a belief which is *too* false will not be useful. But not being too false is consistent with quite a lot of inaccuracy and simplification in our perceptions (e.g. in order to speed up reaction times we might see things as far more sharply delineated than they actually are). Donald Campbell speaks in this context of the usefulness of theories which are 'parsimonious, elegant, (with) few contingencies, few qualifications', by contrast to more complex ones, which are actually closer to the truth, but which may cost too much in time and energy for us to produce in the day to day world of survival and reproduction. We may thus tend to develop perceptions which over-emphasise spatial boundaries and filter out survival irrelevant data, and we could also point to the social utility of a community having beliefs (possibly quite false) which mark its members off from those of other groups. There is the further point that even assuming the broad truth of our evolutionarily produced beliefs and perceptions, an evolutionary explanation will really be applicable only to those beliefs and reaction which will have helped our ancestors get round the savannah and the great plains 10 or 15,000 years ago. All the science which has come later, including the downgrading of secondary qualities, our biology has not prepared us for little or not at all – even leading Niels Bohr to speculate that we are by nature intellectually unsuited to the investigation of the quantum world.

We humans are conscious and self-conscious, and the evolutionary advantages of these traits have been much discussed, rather inconclusively in my opinion. Nevertheless self-consciousness does give us the possibility of scrutinising what we believe, and how we orient ourselves to the world. This sort of scrutiny immediately raises for us the question as to the truth of what we believe, for as we learn from Moore's paradox we cannot realise that we believe something without taking it to be true. So, our long-term evolutionary inheritance has honed our beliefs and perceptions for their usefulness to survival and reproduction, whereas by virtue of our self-consciousness we are also interested in truth. Truth and usefulness may conflict, as already suggested. But more striking is the way that the drive for truth and self-understanding which emerges from our self-consciousness has very little to do with aiding survival. Considering such things as astrophysics, Gödel's theorem, speculative philosophy, poetry and music, and the pre-eminent part they play in the lives of so many of us, we may be tempted to follow Thomas Nagel is asserting that if we 'came to believe that our capacity for objective theory were the product of natural selection, that would warrant serious scepticism about its results beyond a very limited and familiar range', so the development of the human intellect must be seen as 'probably a counter-example to the law that natural selection explains everything' [2].

Neo-Darwinists, such as Geoffrey Miller may concede the point, as far as survival goes, but invoke sexual selection at this point. For Miller, the neo-cortex (the seat of our intellectual activities) is not primarily a survival device at all: it is largely a courtship device to attract and retain sexual mates. So being good at the activities listed above helps us to have sex and to reproduce ourselves. Even if this were true rather more than experience suggests, what we have here is a most an externalist explanation, pointing to the results of being good at poetry, music, physics and the rest: it does nothing to explain why sexual partners are falling over themselves to bed poets, musicians, physicists and the rest (if that is indeed the case). We still need to explain what is so valuable about these and kindred activities, so as to attract the potential mates in the first place (and in doing this, we may well be led into the age old war between Aristotle (as the advocate of contemplation and the pure desire to know) and his utilitarian critics (such as Bacon and Locke) who see science and our knowledge more generally primarily in terms of its potential to improve our world in a technological sense).

My conclusion at this point is to see EE as providing something of a corrective to the unrealistic standpoints and demands of modern epistemology. It is correct and potentially fruitful in seeing us and our intellectual and perceptual faculties as embedded in our biology and reflecting our biological evolution, and in taking our worldly existence as its starting point. To that extent we can defend Lorenz against Kant. On the other hand, the evolutionary model will have difficulties when it attempts to take us beyond the cognitive ambience of the savannah, and it will thoroughly mislead in attempting to analyse all our cognitive and intellectual interests in terms of survival or reproduction, or both.

Thomistic Epistemology (TE)

Perhaps 'Thomistic epistemology' is a misnomer, as Thomas did not see epistemology as the key to philosophy or its starting point. That is the point, though; as with EE, his basic philosophical orientation cuts off a certain type of epistemology before it can get started. We and the world are both created by God, the world as intelligible and we as the potential knowers of it. In a way, where EE has evolving nature as the source of the world we live in and of our cognitive powers, TE has the divine creator as source of both. This move avoids the problems EE is prone to. The whole of creation will be open to our investigations, and our investigations will have a more than utilitarian purpose.

In this perspective, in knowing we, as knowers, fulfil our own intellectual powers. At the same time, the potential intelligibility of the world is realised. Intellectus in actu est intelligibile in actu. In contrast to much modern epistemology and psychology, in TE we (and our minds) are passive, rather than active. 'With us, to understand is in a way to be passive' [3]. This is in part because our intellect cannot be active with regard to everything, yet is has a capacity to understand everything. Thomas quotes Aristotle as saying that the human intellect is like a clean tablet on which nothing is written – but on it everything/anything could be written: '*anima est quodammodo omnia*'.

So, objects we encounter awaken the powers of the soul to understand them. Far from projecting intelligibility on to an otherwise alien and meaningless world, we are from the start at home in the world, belonging to it, formed with and for it (due to the common creation by God of us and world). We participate in the world, and as the mind assimilates the world, it is further assimilated to it. Thomas's view is anti-utilitarian in the following sense: the categories under which we perceive the world are not fundamentally there to reflect and further our interests (as EE might have it). In knowing some object, we perceive its essence, and in bringing its nature into the light, we fulfil its potential. In our contemplation of things and their natures, we are, to a limited degree, participating in God's knowledge.

Questions for TE

TE presents a bracing and challenging contrast to modern epistemology. Nevertheless it seems immediately to raise a number of troubling questions:

1. Is there something particular about human knowing, something only humans could perceive (e.g. time, colour)? Or is this precluded by the clean tablet view?
2. If EE is pessimistic and restrictive in various ways, is TE too optimistic? Can we really know (be) everything? Does its perspective presuppose an unacceptable essentialism, suggesting that we are able in some way to elicit the essences of things? And while all our mental activity may not be to do with survival and reproduction, isn't quite a lot of it rooted in immediate human purposes?

Simone Weil, no utilitarian in a narrow sense, speaks of perception of nature as being a 'sort of dance', based in primitive reflexes and reactions to the world which themselves elicit our perception of the external world. (cf her *Lectures on Philosophy* [4]). So one might accept the participatory side of Aquinas' thinking, while modifying its emphasis both on passivity and on the contemplation of essences.

3. How does the timelessness implicit in the essentialism of TE account for the history of scientific enquiry, and for the picture we are given there of changing and evolving categories and paradigms? How does it count for what looks like a degree of fallibility ineradicable from our researches?

Anthropic Epistemology

To my knowledge, Anthropic Epistemology (AE) does not exist as yet. However, if we take the anthropic principle seriously, it will have epistemological implications, and these will be somewhat between those of EE and TE. Will AE avoid the difficulties of each of these, without generating its own diet of difficult problems?

The anthropic principle rests on taking seriously the high degree of 'fine tuning' there was at the beginning of the universe (or, if there is no beginning, within the universe) for us to be alive and conscious now. Earlier thinkers, such as Betrand Russell and Jacques Monod, emphasised the extreme unlikelihood of any of this (Monod: 'The universe was not pregnant with life, nor the biosphere with man' – [5]). The difficulties consciousness and even life present to a purely physicalistic science are well known to philosophers, and are among the observations exploited by intelligent design theorists.

The fine tuning pointed to by physicists such as Freeman Dyson and Barrow and Tipler does not, of course, take us to an intelligent designer, in the way Intelligent Design Creationism theorists might hope (though it may make us less resistant to entertaining the possibility). But it does suggest that Monod's talk of mankind being a gypsy, living on the boundary of an alien world that is deaf to his music, may be an existentialist exaggeration. If the physical conditions necessary for life were there at the start (or always), in what sense are living creatures in an alien world? If those conditions had to be in an extraordinarily delicate balance, can we really maintain that the pre-biotic universe was not, in a sense, prepared for life? And if those conditions were there at the start, and the universe is huge, spatially and temporally, is it not almost inevitable that life will have occurred some time, and very likely more than once? (And what goes for life here will also go for consciousness.)

According to Paul Davies life and mind do not have to be imported into the physical universe from outside: they 'are etched deeply into the fabric of the cosmos, perhaps through a shadowy, half-glimpsed life principle' [6, pp. 302–303] I should say at this point that I do not really understand Davies' suggestion that this etching is the result of our minds now exercising backward causation on the universe even at the moment of the Big Bang, helping to shape physical reality, as he puts it, even in the far past. [6, p. 287] But I don't see why so extravagant an idea is necessary; mere recognition of the fine tuning could be enough to justify a notion of initial etching. It does not, of course, say what conclusions should be drawn from the point.

Earlier than Davies, Freeman Dyson was happy to say that the universe must have known we were coming [7, p. 250] and he also talked about a universal mind or soul underlying the minds and souls we see around us [7, p. 252]. In *The Anthropic Cosmological Principle* [8], Barrow and Tipler outline a Teilhardian vision of the universe as a whole moving towards an omega point suffused with life and mind, in which the universe will know itself.

As I have just said, we do not need to endorse the more extravagant of these speculations (or move into the strange territory of multiverses) in order to accept that life and consciousness are indeed etched into the fabric of the universe, as the fine tuning point suggests. In *Vital Dust* [9] Christian de Duve extends the argument to show in great detail, (and against the randomness driven account of Monod) how, the importance of chance events notwithstanding in our actual history, the origin and development of life follows an almost inevitable path (see particularly the summary of his argument, [9, pp. 294–300]); if there is fine tuning both at the beginning and during the development of the universe, then it is hardly surprising that our consciousness is able to grasp much about the universe. As Aquinas suggested we will in our very essence be participating in the universe, and are likely to be attuned to its nature, including to its deep structures.

A common reaction to the suggestion that the fine tuning at the beginning of the universe might suggest that life and mind are etched into the fabric of the universe is to say that all the anthropic principle shows is that the conditions necessary for life were present at the Big Bang, and not that those conditions were in any sense sufficient. As a matter of logic, this reaction is correct. Indeed it is more or less a tautology to say that whatever earlier conditions were necessary for some later event must have been present at the earlier time, given that the later event actually took place. This, itself, tells us nothing or very little about the extent to which the later event was, as it were, anticipated in the earlier state.

As I see it, though, the fine tuning point adds to the merely logical point about necessary conditions the further observation that in fact those necessary conditions had to be of an extremely delicately balanced nature. Add to that the extreme difficulty of explaining or even understanding the emergence of life and consciousness from a purely physicalistic point of view, and we may become receptive to the thought that there is be something inherent in the fabric of the universe, revealed partly in those finely tuned initial conditions, which suggests that the universe has a tendency to yield life and consciousness. The primitive dust is vital in de Duve's metaphor.

No doubt this suggestion will be regarded as a throwback to Bergson, an invocation of an élan vital, and it would also be compatible with a Spinozan picture of the relation between mind and matter (though it would not rule out a more traditional theistic view of creation). Many will be inclined to object to it on just these grounds. But rather than ruling it out without giving it a hearing, one could argue that such a perspective makes more sense of our existence in the universe than does the Russell-Monod one, and also makes it more intelligible that we are able to uncover so much about the universe, and in ways that have so little to do with the bare minima necessary for survival and reproduction. At the very least, examining the epistemological implications of the anthropic principle may well be a worthwhile endeavour, as showing how it is that our minds are so attuned to reality at such a variety of levels, and how our aesthetic and moral senses are able to reveal so many unsuspected insights into the universe in ways which appear to us to be utterly compelling (something which, as I argue in my *Beyond Evolution* [10] the theory of evolution has considerable difficulty with).

As EE has it, this is an evolutionary account in a broad sense, because our coming into existence and consciousness has been the result of a universe-long process of physical and biological evolution. So against TE we should expect it to offer an account in which through history we gradually come to know more and better, with the possibility of error ever present. But against EE, there will be no need to confine the validation of our beliefs and mental capacities to those bearing narrowly on survival and reproduction. If we are part of the universe's own long process of development, then so are our thought processes, including our religions and our metaphysical speculations, and our forms of life, including our artistic endeavours. In so far as they need it, they will share in any form of validation accruing to us from our position in the universe's evolution. So there is no need to accede to Darwin's own scruples about accepting as valid only those thought forms we share with the lower animals. If the one is given some evolutionary warrant, so must the other; and if doubt sticks to the higher, then it must to the lower as well, as both have the same source.

AE needs to be developed. But even a cursory glance suggests that it provides a more acceptable starting point than classical epistemology; it shares the world-centered perspective of EE and TE, but may steer a promising middle course between the two.

References

1. Lorenz, K.: Behind the Mirror, p. 9. Metheun, London (1977)
2. Nagel, T.: The View from Nowhere, pp. 79–81. Oxford University Press, Oxford (1986)
3. Aquinas, T.: *Summa Theologiae*. 1.79.2. English translation available at http://www.newadvent.org/summa/
4. Weil, S.: Lectures on Philosophy, p. 52. Cambridge University Press, Cambridge (1978)
5. Monod, J.: Chance and Necessity: An Essay on the Natural Philosophy of Modern Biology, pp. 146–146. Knopf, New York (1971)
6. Davies, P.: The Goldilocks Enigma: Why Is the Universe Just Right for Life? pp. 302–303. Allen Lane, London (2006)
7. Dyson, F.: Disturbing the Universe, p. 250. Harper Row, New York (1979)
8. Barrow, J., Tipler, F.: The Anthropic Cosmological Principle. Oxford University Press, London (1986)
9. de Duve, C.: Vital Dust. Basic Books, New York (1995)
10. O'Hear, A.: Beyond Evolution: Human Nature and the Limits of Evolutionary Explanation. Oxford University Press, Oxford (1997)

Angraecum sesquipedale: Darwin's Great 'Gamble'

Steven Bond

Introduction Karl Popper's 'Recantation'

Canon Charles E. Raven famously minimised the great Darwinian controversy to "a storm in a Victorian tea-cup" [1]. Karl Popper, writing in *The Poverty of Historicism*, added insult to injury by proclaiming that Raven had nevertheless paid too much attention "to the vapours still emerging from the cup" [2, p. 241]. Later, in *Objective Knowledge: An Evolutionary Approach*, Popper elaborated on the perceived 'central problem' of Darwinism as follows:

According to this theory, animals which are not well adapted to their changing environment perish; consequently those which survive (up to a certain moment) must be well adapted. This formula is little short of tautological, because 'for the moment well adapted' means much the same as 'has those qualities which made it survive so far.' In other words, a considerable part of Darwinism is not of the nature of an empirical theory, but is a *logical truism* [2, p. 69].

It may be considered that a formula either is or is not tautological, either it contains empirical content or it does not. And that Popper thus lacks the courage of his conviction in positing that the central problem of Darwinism is its almost tautological status. Given Popper's contention, however, that the 'better' a scientific statement is, the more empirical content it will contain, we must allow for a gradation from tautological statements, through "little short of tautological" statements, to those that are preferable for a riskiness deriving from a yet higher empirical content. A formula's being rejected as "little short of tautological" is at the very least consistent with Popper's philosophical system. And Popper in turn is here consistent with earlier critics in highlighting the tautological nature of evolutionary talk of fitness [3, 4].

S. Bond (✉)
Department of Philosophy, Mary Immaculate College, University of Limerick, Limerick, Ireland
e-mail: Steven.Bond@mic.ul.ie

M. Brinkworth and F. Weinert (eds.), *Evolution 2.0*, The Frontiers Collection,
DOI 10.1007/978-3-642-20496-8_8,

For Popper, the problem with 'survival of the fittest' is that it cannot *but* account for all surviving species by virtue of their fitness, for they have, after all, survived. The pheasant, for example, is well adapted with its feathers camouflaged against high grasslands. One cannot say the same of the male peacock, but his bright tail is nevertheless equally well accounted for by virtue of 'sex selection,' which appears as an '*ad hoc*' addition that explains otherwise falsifying instances of Darwin's core theory. Having survived, every living species is consequently well adapted for survival; the task of the scientist becomes to spot the ingenuity of the adaptations or to admit defeat in the attempt to do so. 'Survival of the fittest' is thus confirmed, but is as devoid of empirical content as a tautology. A falsifying instance in this case would be a surviving species which is not, after all, fit enough to survive. This is a logical impossibility. As such, Darwinism is to be categorised with the pseudo-scientific theories of Freud and Adler, or the historicist theories of Hegel and Marx, and is lacking the explanatory empirical power of Newtonian mechanics or Einstein's relativity. As a theory that makes no risky predictions about the empirical world, it is, in Popper's terminology, unfalsifiable, and consequently pseudo-scientific. Now it would be incorrect to say that Darwin has not made concrete empirical predictions. One entire chapter of the *Origin of Species* was devoted to the 'Imperfection of the Geological Record' and Darwin therein predicts that a more complete record will see the difficulty of our not yet finding numerous transitional links "greatly diminished, or even disappear" [5]. Such a prediction cannot be described as "risky" however, because the Geological Record may ever be deemed imperfect, and we have no time limit on Darwin's either being proven correct or taken to be incorrect. Therefore, while Darwin may be commended for his turn to prediction, it remains unfeasible that the 'incorrect' alternative could ever be the practical outcome of the proposed test. Moreover, a complete fossil record does not equate to a complete catalogue of extinct species, and so the theory of 'transitional links' is evasive enough to survive even a complete and falsifying geological record.

Popper's 'demarcation as falsifiability' gave traditional solace to those seeking a safeguard against the increasing relativity of Kuhn and Feyerabend, and so the former's rejection of Darwinism as 'little short of tautological' received the anticipated backlash of the scientific community. When, in the *New Scientist*, a Dr. Beverly Halstead published an article entitled "Popper: good philosophy, bad science?" Popper himself was quick to support its claims for "the scientific character of the theory of evolution, and of palaeontology" [6]. This is the most famous recantation of Popper's earlier contention that Darwinism lacks predictive capability, though the most eloquent appears in the paper 'Of Clouds and Clocks.' Of the storm in the Victorian teacup, Popper therein confesses "that this cup of tea has become, after all, my cup of tea; and with it I have to eat humble pie" [2, p. 241]. One year previous again he had stated that "The theory of natural selection may be so formulated that it is far from tautological" [7]. It is true also that Popper's description of the growth of knowledge was thenceforth increasingly couched in evolutionary terms, as "the natural selection of hypotheses" for example [2, p. 261]. Nevertheless, the oft repeated contention that Popper was forced to altogether abandon his earlier characterisation of evolution as pseudo-science in the face of stern opposition is to grossly overstate and simplify the case. For the facts remained for Popper that the 'survival of the fittest' could be formulated so as to be "little short of tautological"; that evolutionary theory never succeeds in proposing universal laws the form of which one finds in physics; and that Darwinism needs to be restated if it is not to be altogether abandoned. While evolutionary biology has thus proven itself over time to be a highly successful research programme, we may still criticise Darwin on his own terms, in his own times, without recourse to the later modifications of his theory which others supplied. This is precisely what this paper intends, with specific recourse to the onetime controversial case of *Angraecum sesquipedale*, the Madagascar Star Orchid.

The Curious History of *Angraecum sesquipedale*

Aristide Aubert Du Petit Thouars is a French hero of the Napoleanic wars. When in 1792 his aristocratic birthplace of the castle of Bumois became a target for the rising Reign of Terror, Aristide left in search of the *Astrolabe* and *Boussole*, two ships lost along with their leader – La Pérouse – somewhere in the South Pacific 4 years previous. Subsequent adventures included arrest in Brazil, imprisonment in Lisbon, and exile in the United States where his 12 foot square log cabin in the isolation of Little Loyalsock Creek marked the founding of the town of Dushore. In 1798 he commanded the *Tonnant* at the Battle of the Nile and, as tradition would have it, refused to surrender despite having three of his four limbs removed by cannon balls. Legless, he shouted orders from a bucket of wheat, had the French flag nailed to the mizzen-mast, and was thrown overboard after death according to his own instruction.

A lesser known character is Aristide's older brother Louis-Marie who was detained by revolutionaries in Brest in 1792, and so missed the expedition in search of La Pérouse. Louis-Marie's less adventurous fate was a 2 year imprisonment followed by exile in Madagascar, where he indulged his hobbies of botany and taxonomy by collecting over 2,000 plant species which were later returned to the *Muséum de Paris*. One of the subsequent publications to document this collection was Louis-Marie's 1922 *Historie Particuliére des Plantes Orchidées Recueillies sur les trois Isles Australs d'Afrique*, a work containing the first description of *Angraecum sesquipedale*, more commonly known as the Madagascar Star Orchid. That this orchid is known outside botanical circles today is the singular result of the figurative role it would later play in winning converts to Darwinian evolution.

Darwin had an interest in the cross-fertilisation of orchids since the late 1830s but it notably increased when, in opposition to his doctor's recommendations for rest, he whiled away the heat of July 1861 in the seaside town of Torquay, seeking out orchids and their insect pollinators. On his return to the solitude of Downe, where he had lived and worked now for 20 years, this hobby was continued with renewed vigor. Early morning walks were increasingly diverted by Orchis Bank, where he might catch a fresh glimpse of some curious orchid's contrivance, whilst retaining old hopes of sighting a rare fox as he wandered back home at the dawning. Darwin noted something necessary in the incessant buzzing of bumble bees about red kidney beans in summer; a necessity surmised in his designation of insects as the "Lords of the Floral" [8]. In 1862, 3 years after the publication of *The Origin of Species*, Darwin proceeded with the application of his theory in the lengthy tract, *On the various contrivances by which British and foreign orchids are fertilised by insects*. One Malagasy orchid was of especial interest.

> I must say a few words on the *Angraecum sesquipedale*, of which the large six-rayed flowers, like stars formed of snow-white wax, have excited the admiration of travellers in Madagascar. A whip-like green nectary of astonishing length hangs down beneath the labellum. In several flowers sent me I found the nectaries eleven and a half inches long, with only the lower inch and a half filled with very sweet nectar.....in Madagascar there must be moths with probosces capable of extension to a length of between ten and eleven inches! [9, pp. 197–198].

Darwin proposed a co-evolutionary arms race for ever increasing length of orchid nectaries and moth proboscises, "but the Angraecum has triumphed, for it flourishes and abounds in the forests of Madagascar, and still troubles each moth to insert its proboscis as far as possible in order to drain the last drop of nectar" [9, p. 203]. Darwin speaks of this race with the certainty of one who has watched it trickle on down through the monotonous groan of ages. But while the six rayed Madagascar orchids had long caught the attention of European travellers, no such moth had yet been recorded, nor indeed any insect that was capable of the pollination of this curious flower. Darwin's prediction, consequently, was not very well received.

In 1867, George Campbell (the 8th Duke of Argyll), in an influential book entitled *The Reign of Law*, mocked the 'big noses' of Darwin's predicted moth, describing it as "nothing but the vaguest and most unsatisfactory conjecture" [10]. For it was precisely in those natural curiosities where Darwin found clearest proof of evolution, such as the orchid or the hummingbird, that the Natural Theologians found clearest evidence for the artistry of the creator. In October of this year Darwin writes to Alfred Russell Wallace, whose own work had prompted Darwin to rush publication of *The Origin of Species*, in praise of Wallace's taking up the question of the *Angraecum* in the face of "Duke's attack" [11, p. 281]. While the Duke of Argyll took "the case of *Angraecum* as being necessarily due to the personal contrivance of the Deity" [11, p. 282], Darwin commends Wallace for responding with the question as to why God only imbued things with beauty when it was also functional to do so? No answer was forthcoming. While one may expect a stalemate, however, between an evolutionist and creationist, Darwin also refers in letters to his having been mocked by the entomologists on the issue. Even his close friend Thomas Henry Huxley doubted the various contrivances as recounted by Darwin. Having listened to Darwin describing how the *Catasetum* shoots out its pollinia, Huxley asked simply, "Do you really think I can believe all that?" [11, p. 373]. There are relatively few references to the Star Orchid in the decades that follow the prediction, with Wallace the only one who jumped definitively to Darwin's defence. In the *Quarterly Journal of Science* (1867), Wallace published *Creation by Law*, an essay length book review of Campbell's *The Reign of Law*. Wallace writes,

> That such a moth exists in Madagascar may be safely predicted ; and naturalists who visit that island should search for it with as much confidence as astronomers searched for the planet Neptune, and I venture to predict they will be equally successful! [12].

As it turned out, Wallace's confidence was finally vindicated over 40 years after the publication of Darwin's tract on orchids.

Second Baron Walter Rothschild was a Jewish member of British Parliament, of distant German descent, famous for such idiosyncrasies as keeping kangaroos in his London garden and having his carriage drawn through the streets by a team of African zebra. In 1903, Rothschild co-authored (with Karl Jordon) *A Revision of the Lepidopterous Family Sphingidae*. One new addition to the family bore the elaborate title of *Xanthopan morgani praedicta*. The added 'praedicta' means 'the predicted one', a reference to the fact that Darwin's moth had finally been found. The evolutionists took the discovery of 'the predicted one' as conclusive proof of the theory of evolution itself, and many still do. As recently as November 2004, a *National Geographic* article entitled 'Darwin's Big Idea' invoked the successful prediction as one instance of the "overwhelming" evidence for "Evolution" [13]. It appears a disproportionate leap to make, and is exposed as such on closer inspection.

Wallace's comparison of *Angraecum* to the case of Neptune's discovery is a convenient one, for the example of Neptune has been utilised both in Popper's proclamation of, and in Imre Lakatos's later criticisms of, the theory of 'falsifiability'. It provides a useful case study which highlights some central debates of twentieth century philosophy of science, and so we turn to Neptune to provide us with an interpretative frame through which we might examine more closely the case of the Madagascar Star Orchid.

Karl Popper, Imre Lakatos and the Case of Neptune's Discovery

By 1846, Adams and Leverrier had independently accounted for the residual perturbations in the orbit of Uranus by means of an exterior planet hypothesis. Basing their calculations firmly upon Newtonian mechanics, they independently predicted the position of Neptune to a degree of accuracy which allowed Johann Galle of the Berlin observatory to successfully locate it in September of that year. For Karl Popper, this was a stellar example of scientific method, a series of risky empirically testable predictions which either corroborate (never prove) or falsify the scientific theory at issue. Predicting Neptune's whereabouts to account for the perceived anomalous orbit of Uranus is, on a Popperian reading, a test of Newtonian mechanics, and this openness to falsifiability is what separates Newton's physics from the pseudo-science of historicism or individual psychology.

> For example, Adam's and Leverrier's predictions, which led to the discovery of Neptune, were such a wonderful corroboration of Newton's theory because of the exceeding improbability that an as yet unobserved planet would, by sheer accident, be found in that small region of the sky where their calculations had placed it [14].

Such predictive capability is, indeed, a sight more objectively scientific than Karl Marx's predicting a revolution whilst simultaneously handing out leaflets encouraging the proletariat to revolt. Furthermore, Marxists will always be capable of providing 'ad hoc' alterations in order to maintain their core theory, and so Marxism escapes the falsifiability that the genuinely predictive theories of physics are subjected to. The point is brought home clearly with reference to psychoanalysis in Popper's *Conjectures and Refutations*.

> But what kind of clinical responses would refute to the satisfaction of the analyst, not merely a particular analytic diagnosis but psychoanalysis itself? And have such criteria ever been discussed or agreed upon by analysts? [15].

To use Lakatos's term, the 'hardcore' of psychoanalysis is not under scrutiny. On the contrary, Newton's 'hardcore' (universal gravitation and laws of motion) were under scrutiny, were potentially falsifiable by virtue of the anomalous orbit of Uranus. The latter thus qualifies as science, and the former is reduced to pseudo-science.

In a series of 1973 lectures on scientific method presented at the LSE, Imre Lakatos famously challenged this Popperian bifurcation of theories into genuinely predictive science and the 'ad hoc' revisionism of pseudo-science. For Lakatos, all theories purporting to be scientific have inherited their standards from theology [16 p. 64] and the Newtonian is no less dogmatic in his refusal to overthrow Newton than is the Marxist in the case of Marx. While much of the treatment of Neptune in the 'Lectures on Scientific Method' is not preserved, we do have Lakatos' earlier treatment of same in his 1970 'Falsification and the Methodology of Scientific Research Programmes.'

> A physicist of the pre-Einsteinian era takes Newton's mechanics and his law of gravitation, *N*, the accepted initial conditions, *I*, and calculates, with their help, the path of a newly discovered small planet, *p*. But the planet deviates from the calculated path. Does our Newtonian physicist consider that the deviation is forbidden by Newton's theory and therefore that, once established, it refutes the theory *N*? No [16, p. 68].

Lakatos continues that instead of accepting this 'refutation' of Newton's law of gravitation, yet another new planet (p^1) will be posited, which will require a new telescope in order to be spotted. And if, having constructed the telescope, the predicted planet is not discovered, it will be deemed hidden behind a cloud of cosmic dust. A satellite's being subsequently sent to the site and not detecting the planet will invoke the theory of a magnetic field's interference. And so on and so forth until finally, "Either yet another ingenious auxiliary hypothesis is proposed, or the whole story is buried in the dusty volumes of periodicals, and it is never mentioned again" [16, p. 69]. What Popper refers to as the "wonderful corroboration" of Newton now appears as though it were manufactured as such after the fact, for the 'hardcore' of Newtonian mechanics was not under scrutiny. Lakatos is certainly correct that if Neptune had not been found where predicted, there are plenty of 'auxiliary hypotheses' capable of explaining away this further anomaly, just as Neptune itself was hypothetically invoked to explain the anomaly of Uranus. In the case of an anomaly in the orbit of Uranus, the immediate assumption was not that Newton was incorrect but that some unknown perturbing force will account for the anomaly on strictly Newtonian grounds. Alexis Bouvard, for example, simply rejected the early observations of Uranus as inaccurate, when they could not be fitted into a standard orbit. Thomas Kuhn went yet further towards scientific conventionalism in his treatment of Uranus, noting that despite the likelihood of the planet's being viewed "on at least 17 different occasions between 1690 and 1781," no-one could 'see' it due to its not fitting into the prevailing paradigm of classical astronomy [17, p. 115]. But if we are to take seriously Kuhn's legitimate invocation of 'a role for history' in science, then we cannot follow Kuhn in minimizing the historical fact of Herschel's only 'seeing' the unusual 'disk-size' through "a much improved telescope of his own manufacture" [17, p. 115]. That is to say, there is no need in this instance to invoke a psychological inability to 'see,' when an optical inability circumvents the question. This is not to save Popper from Lakatos's critique, which receives ample illustration here in the tendency of those earlier observers to question the accuracy of their own observations rather than the two millennia old planetary system. The conventionalism of the Kuhnian paradigm may be questionable but Lakatos was nevertheless correct in moving away from the equally questionable anti-conventionalism of Popper.

Karl Popper, however, was notably unreceptive to criticism. In 'Replies to my Critics,' which contains his most lengthy published treatment of Neptune, he clings yet more firmly to a view of science as a series of conjectures and refutations.

> If any of our conjectures goes wrong – if, for example, the planet Uranus does not move exactly as Newton's theory demands-then we have to change the theory. . ..the position of the new planet (Neptune) was calculated, the planet was discovered optically, and it was found that it fully explained the anomalies of Uranus. Thus the auxiliary hypothesis stayed within the Newtonian theoretical framework, and the threatened refutation was transformed into a resounding success [18].

Neptune salvaged a "resounding success" from a "threatened refutation" of Newton's theory – and by this Popper means the 'hardcore' of Newton's theory. But no such threat existed. John Couch Adams stated overtly that 'the law of gravitation was too firmly established' and so could not be called into question 'till every other hypothesis had failed' [19]. Likewise, Johann Galle may have looked through his telescope with a view to giving more credence to Newton (as if it were needed) but he certainly did not look through it with a view to showing Newton up. Of course, given that Neptune was posited as a 'cure' for Uranus not obeying Newton as it ought, the latter is hereby in a win/win situation. An affirmative answer to the question 'Is there a planet precisely here?' will corroborate Newton *further*, but the very posing of the question can be taken as illustration of a general unwillingness to question Newton's core theory. That is to say, the anomalous orbit of Uranus *could* have being conceived as pointing to the inapplicability of Newtonian mechanics to our solar system. Indeed, operating under strict adherence to Popper's theory of demarcation, it *should* have been conceived as such. In practice, what actually occurred was the search for an alternative explanation on Newtonian grounds, the exterior planet only being posited on the continued assumption that Newton was correct. If Newton's 'hard-core' theory really was at issue, at least to the degree that a single falsifying counter-instance is enough to negate the theory upon which empirical predictions are made, then the search for Neptune would never have begun. That it did do so shows the scientific enterprise to be at least partly the attempt to uphold existing theories, and not Popper's perpetual attempt to 'falsify' inherited doctrine.

Lakatos does not put it like this exactly, but points instead to practical illustrations which nicely exemplify the scientific "tenacity" [16, p. 89] against falsification. The choice of terminology shows Lakatos to be swaying towards the epistemological anarchism of Feyerabend, who referred to the scientists' ingenuity in salvaging refuted theories as "the principle of tenacity" [20]. To borrow one of Lakatos's examples, although the proposed perturbing force upon Mercury's anomalous perihelion (the non-existent planet Vulcanus) was not forthcoming; this was no defeat of Newtonian mechanics. The 100 year wait between the 1816 discovery of Mercury's anomaly and Einstein's 1916 explanation of it was not a period in which Newton's theory of universal gravitation was shrouded in doubt, but simply one in which the specific anomaly in question was shelved [16, p. 67]. No experimenter can identify all possible causal influences on the outcome of a given experiment, and is thus without recourse to the ineliminable *ceteris paribus* clause necessary to make any experiment genuinely "risky". Lakatos labels this methodological reluctance to challenge core theories the 'negative heuristic,' an insurance that the 'hardcore' is constantly surrounded by a 'protective belt' of auxiliary hypotheses – whether Newton, Darwin, Freud or Marx be the authors of the 'hardcore' (*not*) in question.

> Popper in *The Logic of Scientific Discovery* believes that the rationality of the scientific enterprise "depends on cutting the propositions into two: basic statements and theoretical statements; falsifiable statements and unfalsifiable statements. This is absolutely crucial because if all theories are unfalsifiable – Popper actually uses the word "metaphysical" to describe them – then Newton and Marx are on a par" [16, p. 90].

Central to Lakatos's criticism is the difference between a merely successful prediction and a genuinely 'risky' one. Had Johann Galle not discovered Neptune where Adams and Le Verrier predicted it, the consequences would be few, if any. We would not have been forced, as Popper so naively proposed, to abandon Newtonian mechanics. Rather, we inheritors of history would not have become acquainted with the names of Adams and Le Verrier at all. Motterlini describes Lakatos's planetary example as designed "to show the link between falsificationism and the '*Duhem-Quine thesis*', according to which 'given sufficient imagination, any theory...can be permanently saved from 'refutation' by some suitable adjustment in the background knowledge in which it is embedded'" [16, p. 68]. And from the perspective of Lakatos, Popperian falsifiability is a step back from the holism of Quine, whose "Two Dogmas of Empiricism" precedes Lakatos in its "blurring of the supposed boundary between speculative metaphysics and natural science" [21]. Against Popper's 'naïve falsifiability', which assumed that Newtonian mechanics was as open to rejection as it was to corroboration, this author sides with the 'sophisticated falsifiability' of Lakatos, which shares Quine's recognition of the scientific unwillingness to readily abandon a core theory in the face of *unsuccessful* empirical applications. Although Popper was slow to admit the relevance of Lakatos's criticism to the case of Newton, he saw its relevance for Darwin readily enough. The very title of Popper's 'Darwinism as a Metaphysical Research Programme' invokes terminology readily identifiable as that used by Lakatos, though it ought not be overlooked that Popper was himself availing of the terminology of MRP's as early as 1949 [22]. Regardless, Popper was aware that MRP's were at this time suggestive of Lakatos's advancements, and his invocation of same points to an acceptance of the 'tenacity' perceived by the latter in relation to this case. Popper, however, doubtless retains his 'hardcore' of falsifiability in relation to scientific methodology more generally when stating that Darwinism "is not a testable scientific theory, but a *metaphysical research programme* – a possible framework for testable scientific theories" [23].

So, in the light of the above discussion, what can we say of the predictive power of Darwinism? Certainly, it is not of the Marxist variety; the great Malagasy moth is no self fulfilling prophecy. And this despite Engel's proclamation at Marx's funeral that "Just as Darwin discovered the law of development of organic nature, so Marx discovered the law of development of human history" [24]. Popper was suitably cautious in referring to Darwinian laws as "*almost* devoid of empirical content" [2, p. 267]. They are not entirely so, and the hawk moth prediction in question offers one clear instance in which the 'survival of the fittest' transcends its *almost* tautological status to provide us with genuine empirical content, and moreover, genuine predictive power. We may now look at this case with the hindsight of the Popper/Lakatos dispute over Neptune in particular, and upon scientific method in general.

But before we cast some doubt upon this Darwinian success story, we ought to say something of the 150+ examples of observable evolution from which this paper does not detract. John Endler's *Natural Selection in the Wild* further infers the probable reasons for said evolution in approximately one third of these cases, including that of wild guppies (*Poecilia reticulate*) researched by Endler himself in the freshwater mountain streams of Venezuela, Trinidad and Tobago [25, 26]. Quantitative comparisons between local populations showed that guppies tend towards drab, well camouflaged colouring in locations of high predation, whilst exhibiting brighter colours and gaudy spots in locations where low predation allows for more highly contested sex selection. Initially, Endler successfully recreated the contrasting conditions in a controlled environment, and within months the guppies successfully tended towards either camouflage or bright colours, as the conditions suggested they might. The continued work of Endler, along with assistance from David Reznick amongst others, has seen guppy populations successfully introduced and observed in the wild with similarly startling results [27]. Progeny become rainbow-coloured where predation is low and camouflage thus of no benefit. And where survival depends upon avoiding predators, males will develop large spots against rocky beds and remain relatively spotless in more uniform, sandier streams. In sum, they evolved precisely as Endler anticipated according to the Darwinian law of the "survival of the fittest."

We might likewise turn to the long nectaries of a species of South African orchid (*Satyrium hallackii*) in illustration of the geographic speciation that Darwin had himself noted. Like *Angraecum sesquipedale*, *Satyrium hallackii* is pollinated by a long-tongued species of hawkmoth. In coastal regions unpopulated by the hawkmoth, however, short-tongued bees have become the pollinator, and the orchids here display the anticipated shorter nectar tubes [28]. Bacterial resistance to drugs provides another prime example of observable evolution, with their short life spans providing ideal fodder for the laboratory study of genetic traits in successive generations. 95% of *Staphylococcus aureus* strains are now resistant to penicillin, compared to 0% when first introduced in 1941 [29]. 'Staph' is likewise evolving resistance to replacement drug methicillin, and a new replacement will be required in the coming decades.

Such a list of empirical evidence for evolution is ever-increasing, and regardless of any doubt shed upon this or that individual example, the fact remains that the evidence for evolution is *in toto* incontrovertible. Nevertheless, this does not detract from the need to engage with specific examples, which are worthy of historical interest in their own right. It is noteworthy that Kettlewell's stock example of industrial melanism in peppered moths, for example [30, 31], which has become the target of much criticism in the last two decades [32–34] consequently receives no mention in two of the most recent book length vindications of evolution [29, 35]. Where an example is no longer utilised to uphold the existing theory, it is typically not used at all. And it is at least desirable that any doubt cast singularly upon Kettlewell's findings will not cast doubt upon Evolution itself. If only to ultimately strengthen Darwin's position, then, by sorting the wheat from the chaff, identifying stock examples which do not provide the level of empirical support they purport to is as necessary as identifying those that do. It is the contention of this author that *Xanthopan morgani praedicta* provides a case in point for the former.

The Case of *Xanthopan morgani praedicta*

> For as well-established a theory as Newtonian physics, it does not look as if falsifiability works like a decisive logical axe that Popper means it to be. And even when a theory is battling against overwhelming odds to establish itself as was Darwin's, metaphysical pigheadedness can prove to be just as recalcitrant [36].

The application of Lakatos' philosophy to the Darwinian prediction at issue yields results not too dissimilar from the case of Neptune above, but there are also some crucial differences. Of similarities, consider that for over 40 years Darwin's prediction remained an unsuccessful one, yet this fact bore no weight in so far as the rejection of Darwinism was concerned. There are the early rejections of Campbell and the entomologists, but at no point throughout the 40 year period of waiting can any critic argue conclusively that the time to wait is over, that Darwin has been proven wrong. The potential protective belt of auxiliary hypotheses was not required on this occasion, though we may yet pose the scenario that it were, and investigate whether in the prediction of the features of an orchid pollinator, there exists such a broad set of auxiliary or saving hypothesis. They are not as readily evident as those which Lakatos highlighted in the case of Neptune, and the prediction of a moth pollinator appears more like a straightforward inference with few saving hypothesis. This apparent divergence of our two examples holds true only insofar as what we are protecting is the hypothesis of a moth pollinator's existence. In the case of Neptune, Lakatos fashioned auxiliary hypotheses designed to save the hypothesis of Neptune's existence from our repeated inability to detect it. It is important to note, however, that it is the 'hard core' of Newtonian mechanics that Lakatos is ultimately engaged in saving. And when, in our chosen example of a moth pollinator, we conceive of the auxiliary hypothesis as designed to save the 'hard core' theory of Evolution, such saving hypothesis are more easily manufactured. It remains a question of one's imaginative power, in conjunction with an experiential knowledge of the case at hand, but the existing literature on *Angraecum sesquipedale* provides us with at least one saving hypothesis for Evolution, supposing that Darwin's proposed co-evolutionary arms-race were ill-founded. In 1997, Wasserthal proposed a 'pollinator shift' model to account for the long spurs of *Angraecum sesquipedale*, which on this account evolved not in conjunction with but subsequent to the long proboscis of the moth. The long proboscis in turn evolved entirely independently of the orchid, and its sole purpose was to allow the moth to escape predation by jumping spiders. Darwin's co-evolutionary race is yet debated by biologists, even given the existence of the 'predicted' one. If, on the other hand, biologists had discovered not the moth predicted but a minute insect pollinator X capable of climbing into the orchid, this too would offer no proof against Evolution. Even an alternative method of propagation does not preclude that the orchid could propagate by more than one method. A saving hypothesis here might be that the proposed co-evolutionary race took place in the distant past before the orchid found a new pollinator X, the moth since having become extinct. Or perhaps our minute pollinator X deliberately seeks deep spurs as a means of protection from predation, thus pollinating only the longest-spurred orchids and leading to their gradually increased length. One could go on fashioning saving hypothesis but the important point remains that the rejection of Darwin's co-evolutionary race will either be for a replacement theory nevertheless consistent with Evolution, or will see the problem 'shelved' until such a time that a replacement theory may be offered.

The tenacity to uphold existing 'paradigms' which Lakatos identified in the case of Neptune finds further reflection in the dearth of criticism Darwin receives in the nineteenth Century scientific journals with reference to the Madagascar Star Orchid. In fact, the few references to *praedicta* which appear between the immediate rejection and *praedicta*'s discovery are either neutral or in Darwin's defence. W. A. Forbes, in the June 12th, 1873 volume of *Nature* asks,

> Can any of your readers tell me whether moths of such a size are known to inhabit Madagascar? They would probably be Sphingidae of some kind, as no other moths would combine sufficient size and length of proboscis [37].

Hermann Müller replied in a brief note entitled *Proboscis Capable of Sucking the Nectar of Angraecum sesquipedale* that his brother had discovered a Brazilian sphinx moth "the proboscis of which has a length of about 0.25 m" [38], a fact to which Darwin referred in the 1877s edition of his orchid tract. The proboscis of the predicted one was certainly not beyond the realm of possibility to late nineteenth century naturalists. Indeed, Wallace's own confidence in the possibility of a suitable moth pollinator was firmly grounded in the fact that he had already witnessed various contenders.

> I have carefully measured the proboscis of a specimen of *Macrosilia cluentius* from South America in the collections of the British Museum, and find it to be nine inches and a quarter long! One from tropical Africa (*Macrosila morganii*) is seven inches and a half [12].

Darwin's great 'gamble' appears less of a Popperian risky prediction once one is told that the Malagasy moth is no natural curiosity, but an advance upon earlier examples of long-proboscis moth species, at least one of which was available to Darwin for scrutiny in the British Museum before the 1862 first edition.

Also, it is worthy of note that Rothschild and Jordon's 1903 *Revision* provides a list of the previous incarnations and nomenclature of *Xanthopan morgani*. Of the four mentioned, the first is Walker's 1856 British Museum specimen *Macrosila morgani*, Sierra Leone and the Congo being its natural habitats. This moth is not only the same species but the very same specimen that Wallace had measured to seven and a half inches in the British Museum. In classifying *morgani praedicta* of Madagascar as a separate subspecies, due to the pinkish tinge of its underside, Rothschild and Jordon are quick to cite Wallace's 1891 *Natural Selection*, in which he reiterated the predicted discovery of Darwin's Malagasy pollinator. Wallace explicitly stated that "two or three inches longer" would do the trick [39, p. 32]. The 'discovery' of Rothschild and Jordon, however, was not the longer proboscis predicted but rather the contestable insight that the outstanding 2 or 3 in. is accounted for by greater quantities of nectar.

> As the tongue of *P. morgani praedicta* is long enough – about 225 mm. = 8 inches – to reach the honey in short and medium-sized nectaries of *Angraecum*, the moths will not abandon the flowers with especially long nectary without trying to reach the fluid, which fills up, in hot-house specimens of *Angraecum*, about one-fourth of the nectar. The result would be that flowers with exceptionally long nectaries would be as well fertilised as such with short nectaries by a moth which could reach the fluid in the long nectaries only when a greater quantity of nectar had collected. *X. morgani praedicta* can do for *Angraecum* what is necessary; we do not believe that there exists in Madagascar a moth with a longer tongue than is found in this Sphingid [39].

Prior to 1903, such proboscises were previously witnessed in Sierra Leone, Congo, Gold Coast, Angola, and were of a genus widespread enough to have its habitat described as "West and East Africa" [39]. What Rothschild and Jordon succeeded in doing was confirming that this moth was also extant in Madagascar. This is a discovery to be sure, and we ought not to take from Darwin's vindication, but nor should we unquestioningly accept the implication of Darwin's having foreseen a curiosity without comparison, an unfathomable 11 in. proboscis that Walter Rothschild miraculously discovered. The 8th Duke of Argyll found it ludicrous, though this ought to be taken as a signpost to his ignorance of exotic flora and fauna, as opposed to that of the century in which he lived. What Rothschild rightly illustrates is the extent to which those who uphold a particular paradigm will do "what is necessary" to uphold it successfully. Earlier in 1903, Francis Darwin, editing a collection of his father's letters, *More Letters of Charles Darwin*, inserted a footnote which attests that "Mr. Forbes has given evidence to show that such an insect does exist in Madagascar" [11, p. 282]. Mr. Forbes, we have seen, has given no such evidence, but prior to Rothschild and Jordon's *Revision*, Müller's response to Forbes was the closest to vindication that Darwin had yet received and so this was all the vindication that was required. As with Müller's "evidence," Rothschild's is open to question. For not only did Rothschild's 8 in. *praedicta* not measure up to Darwin's 11 in. prediction, but it was later determined that the subspecific epithet *praedicta* be withdrawn, as the Malagasy variety was precisely similar to its mainland precursor. It may appear a tad facetious to mention the fact but technically speaking, *X. morgani praedicta* does not exist.

The "sensational victory" of Darwin's little hawk moth, recounted in Part I above and in numerous scientific journals, is based on many real historical instances but is nevertheless a fiction. Gene Kritsky published an article entitled *Darwin's Madagascan hawk moth prediction* in the *American Entomologist*, which opens "on a quite day in January 1862" with Darwin's reception of a package containing Angraecid specimens from Robert Bateman [40, p. 206]. She continues that "so began a 40-year story that illustrated the power of evolution by natural selection...and predicted the existence of a 'gigantic moth,'" but there is little evidence that she grasps the extent to which this really is a 'story' [40, p. 206]. For there was little risk involved in supposing that the "gigantic moth" from the mainland could have found its way to the island, not for the naturalist who as early as 1835 was commenting upon the similarity of Galapagos species to their mainland South American counterparts. Nevertheless, the eventual success of Darwin's prediction saw the name of *Angraecum sesquipedale* revived from obscurity. As with Marx's revolution, or Neptune's discovery, the prediction is used to bolster the relevant core theory, or it is not used at all. The dice, in effect, are already loaded, and a distinctly unscientific picture of *all* historical sciences emerges. Sophisticated and not naïve falsifiability is evident here. Indeed, if Darwin's 'gamble' had not bore fruit, *Angraecum sesquipedale* would likely be little known outside the field of botany today. On this view, Darwin returns to the camp of the Marxist, who endlessly awaits the coming revolution with the fervor of the religious fundamentalist awaiting the Second Coming. The prediction is fulfilled, or one waits a little longer. It is difficult to envision a scenario in which our not having yet located the moth could be utilised as an argument against Darwin, certainly not one which would bear the same force as the vindication Darwin received on *praedicta*'s 'discovery'. What is at issue here is the scope of the prediction which is an existential claim, "There is an orchid pollinator with such-and-such features." Despite being a testable existential claim, its potential vindication is not matched by any apparent openness to falsification. A claim such as "There is a monster in Loch Ness" possesses a definite physical scope in accordance with the limits of the lake itself. Nevertheless, logically, this scope is sufficiently broad to uphold the theoretical possibility of the Loch Ness monster's existence in the face of all failed attempts to precisely locate it. Similarly, the prediction of an "orchid pollinator with such-and-such features" is a prediction of limitless scope, in so far as we do not possess the means to conclusively falsify it. And even in an ideal world where we could view all Angraecids at all times in all places, and determine definitively that no such moth was involved in their propagation, Darwin's minor failure would not impact upon the Metaphysical Research Programme it was designed to support. We can take it as a neat illustration of Lakatos' criticisms, however, that this minor success has been utilised as proof of Evolution itself.

> The significance of this moth prediction goes beyond the historical details. It relates to Darwin's methodology and to his 'evolution by natural selection.' The scientific method dictates that hypotheses are tested by experimentation and that a verified hypothesis takes on the status of a theory. Darwin's experimentation with *sesquipedale* pollination and the confirmation of his moth prediction is entomological verification of the theory of evolution via natural selection [40, p. 209].

Against Kritsky above, an invocation of Lakatos shows that the extent to which evolutionary theory *in Darwin's own time* is making genuinely risky predictions about the world of empirically verifiable facts appears more minimal than first assumed. And Karl Popper's earlier designation of Darwinism as "little short of tautological" is not, based on the specific example in question, very wide of the mark.

The specific example in question is, of course, by no means the end of the issue. Popper was already aware that Darwinism appears empirically empty under a certain formulation only, that it was under the specific formulation of the "survival of the fittest" that it would need to be either abandoned or re-formulated. It is the enterprise of reformulation and not abandonment that has occupied contemporary philosophers and biologists. It is important also to note that talk of the "survival of the fittest," a phrase coined by Herbert Spencer, cannot be found in the first two editions of *The Origin of Species*, which utilise Darwin's own terminology of "natural selection" in its place. Subsequently, Darwin's use of Spencer's formulation gradually increased. As Eliot Sober remarks, it is once the theory is summarized under Spencer's phrase that it is opened to charges of circular reasoning.

> Had he realized the confusions that would ensue, maybe Darwin would have distanced himself from this slogan [41, p. ix].

Whether we yet condone Popper's observation that "a considerable part of Darwinism" is a *logical truism* thus hinges on the extent to which "survival of the fittest" can be said to constitute the considerable part in question. Given that the phrase does not occur in the first two editions of the *Origin*, however, we are reduced here to a critique of Spencer's choice of terminology (and Darwin's endorsement of same), as opposed to highlighting the tautological nature of Darwinism *per se*. Furthermore, more recent attempts to find some non-tautological content in talk of fitness are an advance upon the clear circularity of Spencer's formulation.

Mills and Beatty offer a propensity interpretation of fitness, which they claim "captures the intended reference of 'fitness' as biologists use the term" [42, p. 4]. On this account, attacks of circularity are justified only when defining fitness "in terms of actual survival and reproductive success" [42, p. 5].

> We believe that the confusion involves a misidentification of the *post facto* survival and reproductive success of an organism with the *ability* of an organism to survive and reproduce. We believe that 'fitness' refers to the ability [42, p. 8].

It remains open to contention whether determining 'fitness' as a dispositional property of organisms would yet be influenced by a *post facto* consideration of past performance. Returning momentarily to John Endler's discussion of *Poecilia reticulate*, one might easily anticipate that a brightly coloured male guppy will not be 'fit' to survive predation, but beginning one's research with a quantitative comparison of existing populations in specific environments negates the need for prediction here. Whether design considerations expose one again to charges of circularity is thus dependent on individual practitioners, even granting a dispositional model which presents the possibility of avoiding the *post facto* method. Indeed, Beatty and Finsen (née Mills) later made a convincing argument against their own propensity interpretation [43]. Anticipating Sober, they view the earlier one-generation time scale implied as too short-term, and while further discrepancies emerge in their respective attempts to rectify the problem, the general thrust of Beatty and Finsen, and Sober is consistent: "expected number of offspring is not always the right way to define fitness" [41 p. 26]. It may yet be then that 'fitness' is not tautological, though Spencer's "survival of the fittest" certainly is. This is not to make Popper's critique inaccurate, though its relevance would thus be confined to a criticism of Darwin in his own time and would not impinge upon contemporary formulations of 'fitness' and 'natural selection'.

Conclusion

While it is true that Darwin was led to his prediction through his newly proposed mechanism of 'natural selection,' we may still ask whether the correctness of the prediction necessarily signifies the correctness of the theory. Even the 8th Duke of Argyll had noted of most British orchids an "exact adjustment between the length of its nectary and the proboscis of an insect" [44]. Darwin had done likewise, and one may suppose the supposition of a proboscis to 'fit' the observed nectary to be founded on the habit of constant conjunction, to which an evolutionary explanation is subsequently and arbitrarily attached. That is to say, one may have presumed a 'fit' of proboscis and nectary based on the perceived 'fit' in alternate empirical observations. Indeed, Darwin himself had already witnessed certain species of British Sphinx's collecting nectar via proboscises as long as their bodies. And Darwin's presumption of a longer proboscis yet need not *necessarily* entail the additional assumption of a co-evolutionary arms race. That it did entail such an assumption is quite simply not the issue. For we may yet ask if perhaps an intelligent child might not suppose, on having seen a small flower visited by a short tongued moth, that a big flower must be visited by a long tongued one. Such a prediction, when proven correct, would vindicate neither a theory of evolution, nor the logic of the child. Similarly, if a religious man predicted the great Malagasy moth because he could see no other reason for God to conjure such a curious orchid, the bearing out of this prediction should not be taken as proof of the existence of God. This brings us to one important respect in which the case of Neptune and the hawkmoth differ. Namely, while the prediction of Neptune was impossible without the apparatus of Newton's mechanics, we cannot say the same of the prediction of a hawkmoth in Madagascar. Leaving aside momentarily the debate between Popper and Lakatos as to whether we are testing the 'hardcore' or 'auxiliary hypothesis,' whether our method is 'naïve' or 'sophisticated falsifiability'; we can nevertheless say for certain that the search for Neptune is 'Newtonian' in a concrete and necessary way. The search for a new planet in the sky at large would not necessarily be so, but the search for a planet in a particular patch of sky chosen necessarily upon the assumption of universal gravitation unquestionably is. This is not comparable to the search for a long proboscis in the enormous landmass of Madagascar. In the latter case, there is no necessity attached to the 'Darwinian' nature of this controversy. It is easy to conceive of such a search taking place without any recourse to a theory of evolution.

Of course, a single case study cannot determine the methodology of Darwinism in general, and even if it could it would be an accurate critique of Darwinism only on the questionable assumption of a reliabilist theory of truth. Nevertheless, if we are to unearth the cumbersome facts that lie hidden beneath the clean tales of science's "wonderful corroborations," then we must proceed one tale at a time. It may be that other tales will warrant the title, but the case in point does not. One is at pains not to deny the explanatory power or predictive capability of Darwinism as a Metaphysical Research Programme. Darwin has adequately demonstrated this ability. But when treating specifically of *X. morgani praedicta*, one ought to add that the presumed existence of God possesses the *possibility* of precisely this level of predictive power, and any attempt to salvage an especially scientific predictability in Darwinism requires successful predictions that have been *necessarily* grounded on the truth of evolutionary theory. Their arbitrarily being so is not proof enough.

References

1. Raven, C.E.: Science, Religion, and the Future, p. 33. Cambridge University Press, Cambridge (1943)
2. Popper, K.: Objective Knowledge an Evolutionary Approach, p. 241. Oxford University Press, Oxford (1979) (Revised Edition)
3. Smart, J.J.C.: Philosophy and Scientific Realism. Routledge and Kegan Paul, London (1963)
4. Manser, A.R.: The concept of evolution. Philosophy **40**, 18–34 (1965)
5. Darwin, C.: The Origin of Species, p. 235. Wordsworth Editions, Hertfordshire (1998)
6. Popper, K.: Letter on evolution. A reply to Halstead. New Sci. **87**(1215), 611 (1980)
7. Popper, K.: Natural selection and the emergence of mind. Dialectica **32**, 339–355 (1978)
8. Darwin, F.: The Life of Charles Darwin, p. 303. Tiger Books, Middlesex (1902)
9. Darwin, C.: On the Various Contrivances by Which British and Foreign Orchids Are Fertilised by Insects. John Murray, London (1862)
10. Campbell, G.: The Reign of Law, p. 44. Alexander Strahan, London (1867)
11. Darwin, F., Seward, A.C. (eds.): More Letters of Charles Darwin. A Record of His Work in a Series of Hitherto Unpublished Letters, vol. 1. John Murray, London (1903)
12. Wallace, A.R.: Creation by law. Q. J. Sci. **4**(16), 470–488 (1867), p477n
13. Quammen, D.: Darwin's big idea. Natl. Geogr. **206**(5), 2–35 (2004)
14. Popper, K.: Realism and the aim of science. In: Bartley III, W.W. (ed.) Postscript to the Logic of Scientific Discovery. Hutchinson, London (1983)
15. Popper, K.: Conjectures and Refutations, p. 38. Routledge and Keagan Paul, London (1963)
16. Lakatos, I.: Lectures on scientific method. In: Lakatos, I., Feyerabend, P., Motterlini, M. (eds.) For and Against Method: Including Lakatos's Lectures on Scientific Method and the Lakatos-Feyerabend Correspondence. University of Chicago Press, London (1999)
17. Kuhn, T.S.: The Structure of Scientific Revolutions, 3rd edn, p. 115. University of Chicago Press, London (1996)
18. Popper, K.: Replies to my critics. In: Schilpp, P.A. (ed.) The Philosophy of Karl Popper, bk. 2, pp. 961–1197. Open Court Press, Illinois (1974)
19. Adams, J.C.: An explanation of the observed irregularities in the motion of Uranus, on the hypothesis of disturbances caused by a more distant planet; with a determination of the mass, orbit, and position of the disturbing body. In: Adams, W.G. (ed.) The Scientific Papers of J.C. Adams, vol. 1, p. 7. Cambridge University Press, Cambridge (1896)
20. Feyerabend, P.K.: Consolations for the specialist. In: Musgrave, A., Lakatos, I. (eds.) Criticism and the Growth of Knowledge, p. 205. Cambridge University Press, Cambridge (1970)
21. Quine, W.V.: Two dogmas of empiricism. Philos. Rev. **60**(1), 20 (1951)
22. Popper, K.: Unended Quest: An Intellectual Autobiography, p. 269. Routledge, London (2002) (Updated Routledge Classics Edition)
23. Popper, K.: Darwinism as a metaphysical research programme. In: Rosenberg, A., Balashov, Y. (eds.) Philosophy of Science Contemporary Readings, p. 302. Routledge, London (2002)
24. Engels, F.: Der Sozialdemokrat (Speech made at Karl Marx's funeral, 22 Mar 1883). http://www.marxists.org/archive/marx/works/1883/death/dersoz1.htm. Accessed February 11, 2011
25. Endler, J.A.: Natural selection on color patterns in *Poecilia reticulate*. Evolution **34**, 76–91 (1980)
26. Endler, J.A.: Natural Selection in the Wild. Princeton University Press, Princeton (1986)
27. Reznick, D.N., Shaw, F.H., Rodd, H., Shaw, R.G.: Evaluation of the rate of evolution in natural populations of guppies (*Poecilia reticulate*). Science **275**, 1934–1937 (1997)
28. Johnson, S.D.: Pollination ecotypes of *Satyrium hallackii* (Orchidaceae) in South Africa. Bot. J. Linn. Soc. **123**, 225–235 (1997)
29. Coyne, J.: Why Evolution is True, p. 142. Oxford University Press, Oxford (2009)
30. Kettlewell, H.B.D.: Selection experiments on industrial melanism in the Lepidoptera. Heredity **9**, 323–342 (1955)

31. Kettlewell, H.B.D.: Further selection experiments on industrial melanism in the Lepidoptera. Heredity **10**, 287–301 (1956)
32. Mani, G.S.: Theoretical models of melanism in *Biston betularia* – a review. Biol. J. Linn. Soc. **39**, 355–371 (1990)
33. Berry, R.J.: Industrial melanism and peppered moths (*Biston betularia* [L.]). Biol. J. Linn. Soc. **39**, 301–322 (1990)
34. Wells, J.: Second thoughts about peppered moths. Scientist **13**(11), 13 (1999)
35. Dawkins, R.: The Greatest Show on Earth: The Evidence for Evolution. Free Press, New York (2009)
36. Lee, K.K.: Popper's falsifiability and Darwin's natural selection. Philosophy **44**(170), 291–302 (1969)
37. Forbes, W.A.: Fertilization of orchids. Nature **8**, 121 (1873)
38. Müller, H.: Proboscis capable of sucking the nectar of *Angraecum sesquipedale*. Nature **8**, 223 (1873)
39. Rothschild, W., Jordan, K.: A revision of the *Lepidopterous* family *Sphingidae*. Novitates Zoologica **IX**(supplement), 32 (1903) (Hazell, Watson & Viney, London and Aylesbury)
40. Kritsky, G.: Darwin's Madagascan hawk moth prediction. Am. Entomol. **37**, 206–210 (1991)
41. Sober, E.: The two faces of fitness. In: Sober, E. (ed.) Conceptual Issues in Evolutionary Biology, 3rd edn. MIT Press, London (2006)
42. Mills, K., Beatty, J.: The propensity interpretation of fitness. In: Sober, E. (ed.) Conceptual Issues in Evolutionary Biology, 3rd edn. MIT Press, London (2006)
43. Beatty, J., Finsen, S.: Rethinking the propensity interpretation – a peek inside Pandora's box. In: Ruse, M. (ed.) What the Philosophy of Biology Is, pp. 17–30. Kluwer, Dordrecht (1989)
44. Campbell, G.: The Reign of Law, 4th edn, p. 46. Routledge & Sons, New York (1873)

Darwinian Inferences

Robert Nola and Friedel Weinert

Introduction: Hypothetico-Deductive (HD) Scheme

There have been some attempts in the literature to associate Darwin's reasoning with the hypothetico-deductive methodology ([1] Chap. 1; [2], p. 198). Popper's method of falsificationism is such a hypothetico-deductive procedure and the question arises whether Popper's method adequately characterizes Darwin's method in *The Origin of Species*. According to Karl Popper, science progresses by the method of falsification:

Let theory, T, entail prediction p: $T \rightarrow p$. Let the prediction, p, be found to be falsified, $\neg p$, then the whole theory is falsified, by *Modus Tollens*:

$$[(T \rightarrow p) \wedge \neg p] \rightarrow \neg T].$$

This simple scheme can be made more realistic by adding auxiliary assumption A, such that $[(T\&A) \rightarrow p]$ but this leads to a consideration of the Duhem-Quine thesis, which will not occupy us here. Popper assumes that a theory is universal and has deductive consequences, especially in the form of *novel* predictions. As Popper accepts Hume's criticism of (enumerative) induction, he stresses the asymmetry between verification and falsification. A universal theory can never be conclusively verified but it can be shown to be in conflict with an observational claim:

$$\forall x(Vx \supset Ox); \exists x(Vx \wedge \neg Ox).$$

R. Nola (✉)
Department of Philosophy, University of Auckland, Auckland, New Zealand
e-mail: r.nola@auckland.ac.nz

F. Weinert
Division of Humanities, University of Bradford, Bradford BD7 1DP, UK
e-mail: f.weinert@brad.ac.uk

M. Brinkworth and F. Weinert (eds.), *Evolution 2.0*, The Frontiers Collection,
DOI 10.1007/978-3-642-20496-8_9,

The language of symbolic logic shows well which logical forms need to be adopted for a universal theory to be falsifiable. For a universal statement, like ‘All vertebrates are omnivores’, $\forall x(Vx \supset Ox)$, to be falsifiable, and hence empirical on Popper’s criterion, there needs to be a potential falsifier, viz. some vertebrates which are not omnivores: $\exists x(Vx \wedge \neg Ox)$.

Whilst Popper’s theory may work well for some deductive theories, it fails to reflect the richness of cases in the history of science. Also many scientific theories do not make strikingly novel predictions but accommodate the already known evidence. This is certainly true for the theory of natural selection; Darwin found evidence in breeding practices and fossil evidence, dating back to the 1840s.

Popper’s criterion also makes some theories ‘scientific’, which would on other criteria not be regarded as scientific. All that his criterion requires is that a theory, to be scientific, must be cast in such a form and that it has testable consequences [3, §2.8]. But many of the theories, which Popper criticizes (Marxism, Freudianism) can be cast in forms in which they have testable consequences. For instance, it follows from the theory of Intelligent Design (ID) that vertebrates have complex eyes, although this is not a novel prediction.[1] As this statement is a deductive consequence of ID, the theory of Intelligent Design seems to be testable, by Popper’s criterion. This consequence is clearly not what Popper had in mind for his demarcation criterion. Popper required that the deductive consequences be empirical in nature so that they are objective and intersubjective. These criteria are satisfied by the above statement, but they render ID scientific. However, if we amend the above-mentioned statement to ‘This vertebrate eye evolved, according to a pre-existing design’, we make the deductive consequence untestable. Whilst it can be shown that eyes did evolve in a convergent manner, it cannot be shown that they evolved according to a pre-existing plan. In the case of ID Popper’s criterion does not seem to be very reliable. In this paper we will not be concerned with the demarcation criterion but with the question of how scientific theories compare with respect to their explanatory import in the face of available evidence. In order to do so we will consider a certain type of inference to the mostly likely explanation using the language of probability. (This type of inference is sometimes known as a form of eliminative induction.)

It is sometimes alleged that Popper’s criterion fails for statistical theories because any number of exceptions seem to be compatible with a statistical average. ([5], §3.5; [6], part III) The evolutionary theory is statistical in nature: its observational consequences only follow with a certain probability. This means that the observation of a negative instance is compatible, to a certain degree, with the universal, statistical theory. How many counter-instances does it take to make the theory incompatible with the evidence? Contrary to what is sometimes said, this is not necessarily a strong objection against falsificationism. The half-life of a

[1] At least it is not a novel prediction in the sense of not having been known before the theory was constructed. But Lakatos and Zahar ([4], Chap. 4) have a weaker notion in which a fact is novel for a theory if it was not used in the construction of the theory, but was known before.

particular ensemble of atoms is given by the Rutherford-Soddy formula: $N_o = N_t e^{-\gamma t}$ which can easily be falsified. For instance, for thorium X the half-life is 3.64 days and this figure can readily be tested in the laboratory. But such testing may be more difficult in cases, like Darwinism, where the statistical statement is less precise, because it is then more difficult to define a tolerable level of exceptions. Again it is desirable to consider an alternative approach.

This is not to say that the theory of natural selection does not have falsifiability conditions. In Chap. 6 entitled 'Difficulties on Theory' Darwin tells us that he is aware of the conditions under which his own theory would be falsified if they were to arise:

> If it could be demonstrated that any complex organ existed, which could not possibly have been formed by numerous, successive, slight modifications, my theory would absolutely break down. But I can find no case. No doubt many organs exist of which we do not know the transitional grades. ... We should be extremely cautious in concluding that an organ could not have been formed by transitional grades of some kind [7, pp. 190–191] (190–191, 1st edition online).

The potential falsifier for his theory is that there exists some complex organ which did not arise by successive slight modifications of an earlier form of organ. Merely not knowing what the earlier form was, or lacking evidence for it, would not be sufficient to turn the potential falsifier into an actual falsifier; what is needed is that there actually be some organ that has no earlier form from which it arose by slight modification. Establishing that there be such an actual falsifier is, of course, not a straightforward matter.

From the point of view of present considerations, the most serious problem with the method of falsification is that it constructs a confrontation between one particular theory and the evidence it predicts, as reflected in the scheme $[(T \rightarrow p) \wedge \neg p)] \rightarrow \neg T]$. Whilst this scheme may fit some specific examples, it does not seem to reflect many cases in the history of science where we usually see a pair of rival theories, T_1 & T_2, claiming to explain the evidence equally well, where the evidence does not follow deductively from T_1 or T_2.

Despite Darwin's acknowledgement of falsifiability, the aim of this paper is to show that Darwin employed inferential practices, which are not hypothetico-deductive; rather they require a comparison of his own theory of Natural Selection with rival theories.

For instance Darwin opposed his own theory of evolution to the then popular theory of natural theology; today Darwinism is opposed to Intelligent Design. If that is the case why do commentators on Darwinism, like Ghiselin and Ruse, associate Darwin's procedure with the hypothetico-deductive method? For instance, according to M. Ruse, Darwin wanted to make 'his theory as Newtonian as possible' [2, p. 176] in the sense that it can be set out as a system of fundamental principles from which phenomena can be derived. M. Ghiselin sees strong similarities between Darwin's hypothetico-deductive method and Popper's falsificationism [1, p. 5]. Ghiselin's insistence on Darwin's hypothetico-deductive method goes hand in hand with the rejection of 'Baconian induction'. Although Darwin paid lip service to 'Baconian induction' [1, p. 35], his theory was actually based 'upon the construction of hypothetico-deductive systems' [1, p. 63]. However, this argument is erroneous from two points of view. Popper's HD scheme requires the prediction of *novel* facts, which if successful, increases the corroboration of the theory under test, whilst Darwin's work consisted mostly in the accommodation of *known* facts. Popper's HD scheme is deductive in nature but Darwin's theory is statistical and hence has probabilistic consequences. It is of course possible to present Darwin's system as a deductive system, once the principles of his theory are in place and use deductive consequences as confirmation of the principles [2, p. 62]. But this is not how Darwin proceeded in *The Origin of Species*. It is true that he pays lip service to Baconian induction, but the second mistake is to think that Baconian induction is induction by enumeration [1, p. 230]. It is in fact induction by elimination or at least some form of Inference to the Best Explanation (IBE), to be discussed below. Bacon explains the difference in the following passage:

> For the induction which proceeds by simple enumeration is childish; its conclusions are precarious, and exposed to peril from a contradictory instance; and it generally decides on too small a number of facts, and on those only which are at hand. But the induction which is to be available for the discovery and demonstration of the sciences and arts must analyse nature by proper rejections and exclusions; and then, after a sufficient number of negatives, come to a conclusion on the affirmative instances...([8], Book I, §105)

Induction by elimination may be compared to the work of a detective, who attempts to solve a crime. The potential suspects are matched against the available evidence. Those whose profile is incompatible with the evidence are eliminated. Thus, if the crime happened at location A, but one suspect was at location B at the time of the crime, this suspect is eliminated. To reflect these facts about the history of science in general and Darwin's theory in particular, it is necessary (a) to distinguish various forms of inferences and (b) to apply these inferential practices to some of Darwin's case studies.

Some Forms of Inference to the Best Explanation (IBE)

A number of writers note the role that some form of Inference to the Best Explanation (IBE) plays in Darwin's thinking for example Kitcher, ([9], Chap. 3, pp. 43–58) and Lewens (10], Chap. 4). Lewens claims that appeals to IBE can sometimes be sloganistic and ought to be more than a mere appeal to likelihoods. In a number of publications Elliott Sober advocates what he calls a Law of Likelihood which is strictly not IBE on some of its standard interpretations ([3], Sect. 1.3). The task of this section will be to explore some of the various forms IBE can take in characterising Darwin's arguments.

The premises of an IBE argument begin with a set of facts F and a (finite) number of candidate explanations $\{T_1, T_2, .., T_n\}$ of these facts. What is the task of the inference form IBE? If it is to find the actual explanation of F then IBE will fail if the actual explanation is not in the initial set of explanations. A less ambitious task would be to find the best explanation in the set, or the most acceptable or the most favoured. Expressed schematically in the case of two candidate explanations, IBE is a non-deductive argument of the following sort:

1. F
2. T_1 and T_2 are two rival candidate explanations of F
3. T_1 is a better explainer than T_2 of F
4. ∴ ???

There are a number of candidate conclusions. (a) The first is that the better explainer, T_1, is true. Since few think that IBE is such a reliable form of inference[2] we will not pursue this here. (b) In Peirce's characterisation of abduction ([12], Chap. 11) the conclusion contains an epistemic operator in front yielding a conclusion such as 'it is reasonable to accept T_1 as true'. This is much weaker than conclusion (a) and allows that even if it is reasonable to accept that T_1 is true that T_1 might actually be false.[3] (c) Yet others recommend a weaker conclusion than (b) such as 'T_1 is to be *accepted* over T_2', i.e., the bare acceptance of a theory rather than its being *reasonable* to accept. (d) The final conclusion to be considered claims that 'T_1 is to be *favoured* over T_2'. Favouring is a contrastive notion favoured by those who put emphasis on the notion of likelihood in explanations. Conclusion (d) comes close to the conclusion Elliott Sober wishes to draw on the basis of his version of IBE which uses the Law of Likelihood. For this reason some may insist that this is not strictly a form of IBE at all. To mark the difference we could call this IMF – *Inference to the More Favoured.*

What about the premises? Premise (1) concerning some fact or set of facts is unproblematic – it is just a given. Premise (2) concerns the set of candidate explanations that are at hand. As already noted this might not contain the correct explanation; in respect of both T_1 and T_2 the truth may lie elsewhere in some further theory perhaps not even envisaged or formulated. One way of overcoming this problem is to employ a third "catch-all" hypothesis: (neither T_1 nor T_2). How the set of candidate explanatory hypotheses are to be specified is not a straight-forward matter that needs to be addressed in a full account of IBE; but this does not concern IMF since it merely considers any pair of hypotheses and determines which is more favoured by facts F.

Finally some might require an additional premise to the effect that the two explanatory hypotheses meet minimal standards for being satisfactory explanations; otherwise the claim that T_1 is a better explainer than T_2 allows that T_1 is the best of a bad bunch. But in some cases this is in fact what is uncovered when considering a pair of rival explanations. Darwin claimed that his

[2]Niiniluoto [11] in his account of IBE argues for a more complex relation between explanation and truth in showing that there is a link between approximate explanatory success and truthlikeness.

[3]Musgrave ([13], Chap. 14) construes IBE differently as a deductive argument but it has a conclusion like (b). He also argues that the claim that it is reasonable to accept T_1 is consistent with T_1 being false.

explanatory hypothesis of natural selection offered good explanations while the rival explanation of creationism offered none: 'On the ordinary view of the independent creation of each being, we can only say that so it is; – that it has so pleased the Creator to construct each animal and plant' ([7] (Darwin 1859–1872, 435, 1st edition online)). In Darwin's view creationism offers no explanation at all as to why some creature possesses some feature. In the case of IBE the 'best of a bad bunch' objection has some force. But since IMF is not as ambitious as IBE in that it seeks the more favoured of a pair of hypotheses, this objection has little weight.

There is a large literature on different models of explanation (Hempelian deductive nomological, causal, etc.); in general we can remain neutral as to which model we adopt in our schematic account of the forms of IBE. However we will adopt an account of explanation which at its core can be expressed in terms of likelihoods. In contrast there is a much smaller literature on how explanations are to be ordered, that is, how one determines when one explanation is better or worse than another, a matter which is crucial for premise (3). This matter can be addressed if explanations are understood as likelihoods.

Can we cash out the notion of explanation in terms of the notion of likelihood? And can we take offering a *better* explanation as having a *higher* likelihood? If so we would have an account of explanation in terms of likelihoods and an account of when one theory offers a *better* explanation of the same facts F in terms of *higher* likelihoods. One proposal for an account of what a better explanation in terms of a comparison of likelihoods can be set out along the following lines:

(i) For two candidate explanatory theories T_1 and T_2, and some fact F, T_1 is a *better explanation* of F than $T_2 =_{\mathrm{Defn}} \mathrm{prob}(F, T_1) > \mathrm{prob}(F, T_2)$.

Certainly if T_1 explains fact(s) F, and it explains F better than T_2 does, then at least T_1 makes F more probable than T_2 does; that is we have $p(F, T_1) > p(F, T_2)$. But there are counterexamples to the definition just proposed such as Sober's "gremlin" hypothesis (see Sober [14], Sect. 2.2). Suppose one hears a rumbling in the attic (fact R). One potential explanation would be the hypothesis G: there are gremlins in the attic who are using it as a bowling alley. In this example it can be the case that the likelihood prob(R, G) is high; bowling gremlins do make noises such as those heard coming from the attic. Moreover the G hypothesis can make R as highly probable as any other hypothesis one might entertain. Though G makes the fact R highly likely, G is defective in other ways. What evidence is there for the gremlin hypothesis G? Hardly any. If we were to ask, 'what is the probability of G on the basis of any other evidence E outside the context in which G is to be explained (viz., prob(G, E))?', then we would have to admit that prob (G, E) was low or zero (since there is much evidence against the existence of gremlins in the first place). So being a better explanation is not always just having a high likelihood; in many cases one has to take into account such prior evidential probability of the hypotheses as well. But once we have done this we can think of this as just the prior probability prob(G), that is, the relative prior probability of G prior to the use of G to explain fact R.

To accommodate this point, a modification can be made to the above definition:

(ii) For two potential explanatory theories T_1 and T_2, and some fact F, T_1 is a *better explanation* of F than $T_2 =_{\text{Defn}}$ (1) prob(F, T_1) > prob(F, T_2) (comparison of likelihoods); and (2) prob(T_1) > prob(T_2) (comparison of prior probabilities).

Granted this, we can understand explanation in the schema for IBE to employ the above notion of better explanation. And we can understand IMF to also employ the above notion of better explanation where the conclusion is (d) above, viz., T_1 is to be favoured over T_2.[4]

This is akin to an account of explanation, and better explanation, recommended by van Fraassen [16, p. 22]. He mentions statistical practices which quite often use only clause (1). Clause (2) turns on an account of initial or prior probabilities which tell us what is the initial plausibility of each theory when used just in the context of explaining fact(s) F, as in the case of the "gremlin" hypothesis. Such prior probabilities are only "relatively" prior in that other background evidence may well have had to have been taken into account in an independent assessment of the plausibility of the hypotheses. Other considerations can have a bearing on relative prior probability such as consilience, as is illustrated in the example drawn from Darwin discussed in section "Blind Cave Insects".

The above does not show that IMF is a form of inference to be avoided. Rather it displays the way in which IMF is to be understood; it turns on the notion of the *favouring* of one of a pair of hypotheses, or of *differentially supporting* one of a pair of hypotheses, rather than offering non-comparative support *tout court* for just one hypothesis. As we will see, Darwin often makes his points in favour of Natural Selection and against Special Creationism in just this way. In sum, we can say that IMF is the following inference involving contrastive explanation:

1. F
2. T_1 and T_2 are two rival candidate explanations of F
3. T_1 is a better explainer than T_2 (as understood in (ii) above)
4. $\therefore$ T_1 is to be favoured over T_2.

The above takes into account not just the likelihood of the explanatory theory but also its probability on available evidence. Where can this probabilistic support come from? In some cases it may be evidence gathered from prior investigations, as is illustrated in the case of Sober's gremlins. An important role can also be played by plausibility considerations in determining values for the (relative) prior probabilities of explanatory hypotheses (see [17], Chaps. 4 and 18, see section "Intelligent Design Versus Evolution" below). But further considerations arising from consilience also feed into how we are to assess relative priors.

---

[4]Something like the above is adopted as an Explanation Ranking Condition in Glass ([15], p. 282). The idea here is that an account of a better explanation needs to give the same result as clauses (1) and (2) do when they give the same result. Glass argues that they need not give the same result and he recommends a coherence measure for better explanation, a matter not discussed here.

Consilience is a Latin-based term introduced by William Whewell literally meaning "jumping together". Whewell's own account of consilience says that extra confirmatory support accrues to a theory which is initially constructed to explain one class of fact and is then discovered to explain other independent classes of fact. The independent classes of fact brought under the scope of the one theory obtain a unity that they would otherwise lack in the absence of the theory. Whewell also argues that consilience shows that the theory has the stamp of truth. (See [18], p. 295). But that a theory "jumps together" two or more disparate facts does not show that the theory is true; false theories can also do this. So this is a claim of Whewell's that we can set aside.

A classic example of consilience is that of Kepler's three laws of motion (which are logically independent of one another) with the introduction of Newton's Law of Universal Gravitation. Newton's Law was constructed by Newton on the assumption of Kepler's third law (the constant proportionality of the cube of the mean distance of a planet from the Sun to the square of the periodic time taken to orbit the Sun). Newton then showed that his Law entailed Kepler's other two laws. Thus Newton's Law, in consiliating Kepler's three laws and other independent laws such as Galileo's free fall laws, gives these laws a unity that they would otherwise lack. In accounting for such independent, additional law-like facts for which it was not originally constructed, Newton's law also obtains extra confirmation from those facts. More generally we can say that consilience occurs when two or more independent classes of fact (including laws) are "jumped together" by a theory that would, in the absence of the theory, remain independent. Here consilience is like unification in that it brings disparate facts within the scope of one theory.

We can expand on this in the following way. Consider a pair of rival theories T_1 and T_2 in which T_1 "jumps together" a number of independent facts F while T_2 fails to "jump together" any of the facts F. Can we say that T_1 gets extra confirmational support by F above that of a rival T_2? Intuitively one might think that this is so. And this is correct since there is a proof of this given in McGrew [19] which can be summed up as follows:

> the degree of confirmation a hypothesis receives from the conjunction of independent pieces of evidence is a monotonic function of the extent to which those pieces of evidence can be seen to be positively relevant to one another in the light of that hypothesis ([19], p. 562).

This result provides an important link between the consilience a theory can produce of facts F and the extra confirmation that accrues to the theory from this. It is an important consideration to take into account in the comparison of Darwin's theory of Natural Selection with any rival such as Special Creationism. As will be seen, Natural Selection conciliates a number of biological facts while Creationism fails to conciliate any of them and leaves them as independent facts. In virtue of this Natural Selection gets extra confirmatory support over and above whatever support Creationism has (which may be little enough). Darwin would not have known of this result in any explicit way. But it is one way of understanding what Darwin is getting at when he compares his theory of Natural Selection with Creationism and finds Creationism seriously wanting as a potential explanation of these facts.

Some Applications of the Methodological Principles

This section serves to show how Darwin applied the inferential practices, outlined above, in his work. The section "Intelligent Design Versus Evolution" discusses the role of explanatory mechanisms in rival hypotheses and the section "Blind Cave Insects" discusses the work of consilience and a form of inference to the best explanation, viz., inference to the more favoured explanation, in the case of blind cave insects.

Intelligent Design Versus Evolution

In the history of Darwinism, two theories with incompatible principles (natural selection *versus* design) claim to be compatible with the evidence and to explain the evidence better than their respective rival. In Darwin's time, the evidence derived from comparative anatomy, embryology and palaeontology (Fig. 1a–c).

The recent debate between the Intelligent Design proponent and the evolutionary biologist has often centred on complex organs, like the eye. In terms of explanation as a comparative enterprise, it involves claims of the following kind:

Claim$_1$. One hypothesis, H_1, makes the evidence O more probable than the rival hypothesis, H_2:

$$\text{prob}(O|H_1) > \text{prob}(O|H_2).$$

Claim$_1$ can either be cast as a claim supported by evolutionary biologists, in which case O = eye; H_1 = evolutionary theory; H_2 = Intelligent design theory; or as a claim of the ID theorist, in which case O = eye; H_1 = design theory; H_2 = evolutionary theory.

Consider, first, the Intelligent Design theory. It claims that

$$\text{prob(the human eye has features } F_1..F_n|\text{Evolution) is small}$$

but

$$\text{prob(the human eye has features } F_1 \ldots F_n|\text{ID) is larger.}$$

However, this simple claim is not very realistic, since it can be made easily but the probabilities are difficult to assess: a completely false theory can account for large bodies of fact. The classical case is presented by geocentrism, developed by Aristotle and Ptolemy. Roughly speaking, geocentrism models the universe with the Earth at its geometric centre; the Earth is completely stationary thus possesses neither a diurnal nor an annual motion. The six planets, as they were known to antiquity, circle the central Earth in circular orbs and the sun occupies the orbit, which is now occupied by the Earth. Although this model of the universe is a

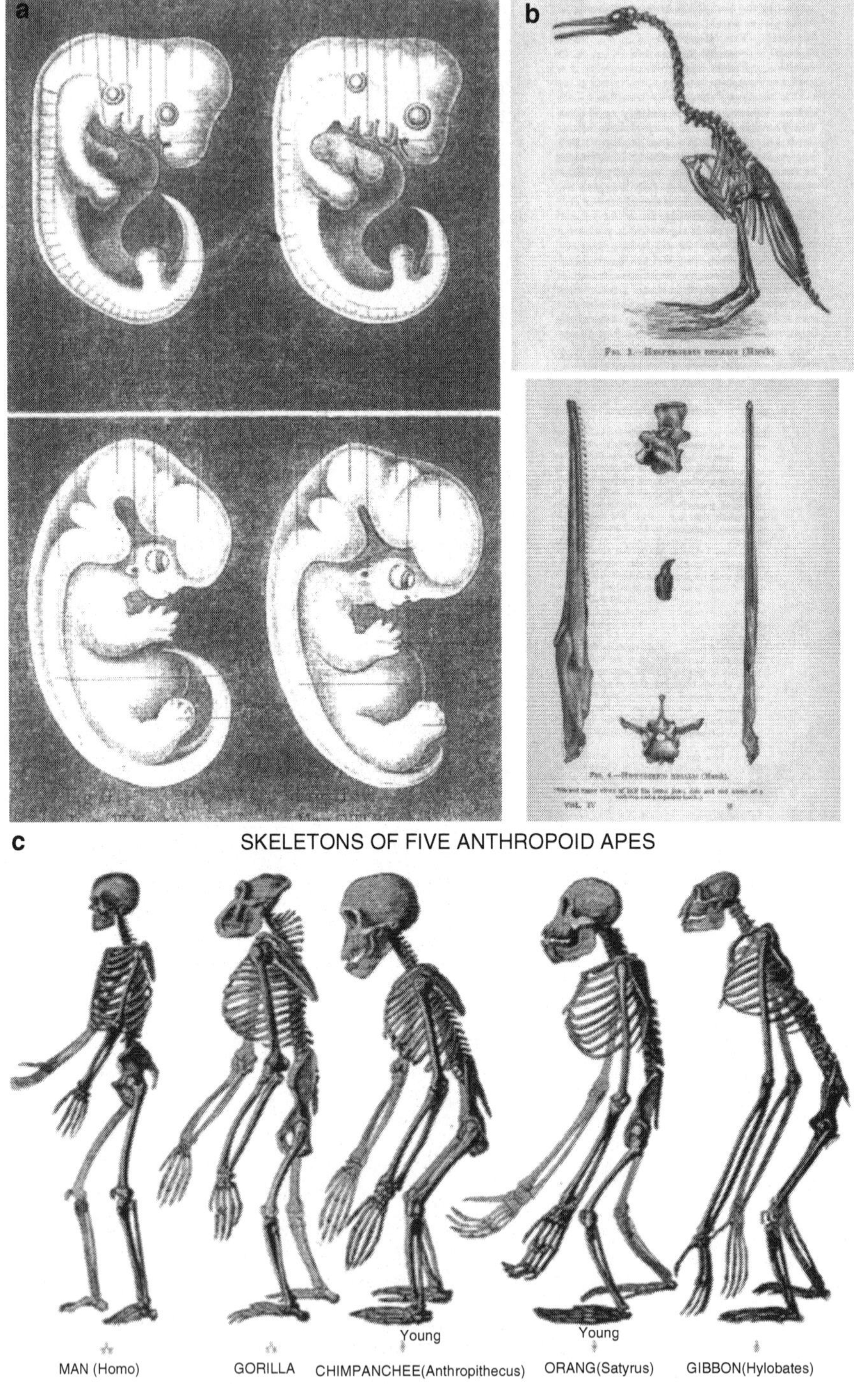

Fig. 1 (**a**) *Embryology*. Top shows a dog (4 weeks old, *left*) and a human embryo (4 weeks old, *right*); bottom shows a dog (6 weeks old, *left*) and a human (8 weeks old, *right*) (**b**) *Palaeontology*. Hesperornis Regalis was an intermediary species between dinosaur and bird. It had both wings and teeth. The bottom picture shows details of the creature's jaws and teeth. (**c**) *Anatomy*. A comparison of skeletons (*left* to *right*) human, gorilla, chimpanzee, orang and gibbon

fundamental misrepresentation of the reality of the solar system, geocentrism was nevertheless able to predict the position of each planet to an accuracy of 5% of modern values. Hence it is important to remember that scientific theories must be explanatory; they must solve genuine problems by objective techniques. Prediction is not sufficient. In their attempt to explain various phenomena scientific theories often explain the evidence by appeal to testable mechanisms. In view of these remarks Claim_1 should be reformulated to become Claim_2:

Claim_2. One hypothesis, h_1 *and* its explanatory mechanism, M_1, make the evidence more probable than its rival hypothesis, h_2, *and* its mechanism, M_2. From the point of view of evolutionary theory, claim_2 states:

$$\text{prob}(O \mid H_1 \,\&\, M_1) > \text{prob}\,(O \mid H_2 \,\&\, M_2);$$

O = eye; H_1 = evolutionary theory; H_2 = Intelligent design theory; M_1 = natural selection; M_2 = separate acts of creation.

The reason for the greater probability conferred on the evolutionary theory is that acts of intelligent design have no testable consequences, but natural selection is a testable mechanism. The work of natural selection can be studied under controlled conditions but an act of intelligent design does not represent a testable mechanism. This latter claim falls outside the realm of science.

In *The Origin of Species* [7] Darwin developed probability arguments against the arguments of Natural theology which favours design. Darwin asked how likely the evidence was in the face of the theory of natural selection as opposed to the theory of special creation. Darwin needed to show that to account for the diversity of species, separate acts of creation were less likely than the principle of natural selection. At Darwin's time there was more general consensus on the facts of evolution than on the underlying mechanism. Lamarck, for instance, had proposed his theory of use inheritance (1809), which Darwin rejected in favour of natural selection. Darwin needed to convince his readers that natural selection was a plausible mechanism, which could elegantly account for many observations. In *The Origin of Species* Darwin repeatedly appeals to probability considerations to argue against the theory of special creations and in favour of natural selection. Darwin argues that a naturalistic process like natural selection is more probable, renders the evidence more coherent and the evolutionary theory more plausible.

> If then we have under nature variability and a powerful agent always ready to act and select, why should we doubt that variation in any way useful to beings, under their excessively complex relations of life, would be preserved, accumulated, and inherited? (...) What limit can we put to this power, acting during long ages and rigidly scrutinising the whole constitution, structure and habits of each creature, – favouring the good and rejecting the bad? I can see no limit to this power, in slowly and beautifully adapting each form to the most complex relations of life. The theory of natural selection, even if we looked no further than this, seems to me to be in itself probable ([7], 1st edition).

Still, as there was at Darwin's time no direct evidence for natural selection, he employs a battery of facts, which on the one hand tend to 'corroborate' the probability of natural selection as a force in nature [7, p. 263], whilst on the other hand they tend to discredit a process like design as improbable. These facts can conveniently be summarized under the aspects of the biodiversity of species, the extinction of old and the emergence of new species and the affinities between species (homologies and analogies). While the theory of natural selection provides a coherent explanation of these facts Darwin exclaims: 'How inexplicable are these facts on the ordinary view of creation!' [7, p. 437]. And he adds: 'He who rejects it, rejects the *vera causa* of ordinary generation ... and calls in the agency of a miracle'. [7, p. 352]

In the 6th edition of the *Origin* Darwin uses the analogy with other theories to further support his theory of natural selection.

> It can hardly be supposed that a false theory would explain, in so satisfactory a manner as does the theory of natural selection, for several large classes of facts above specified. It has recently been objected that this is an unsafe method of arguing: but it is a method used in judging common events of life, and has often been used by the greatest natural philosophers. The undulatory theory of light has thus been arrived at; and the belief of the revolution of the Earth on its own axis was until lately supported by hardly any direct evidence ([7], p. 421 6th ed. online).

Darwin's claim that 'a false theory' could not explain satisfactorily so many diverse facts is misleading, because the geocentric model seemed to explain 'several large classes of facts', although it was entirely false, since it postulated that the Earth was at the 'centre' of the universe. The fact that the evolutionary theory seemed to explain coherently so many facts should hardly be a recommendation for its 'truth'. Strictly speaking, the argument is about the likelihood of the rival hypotheses in the face of the known evidence. It means that the evidence assigns differential probability weights to the contrasting explanations. This procedure is particularly effective, if the rival models do not just face the evidence but if the models are enhanced by a specification of the underlying mechanism, which is supposed to be causally responsible for the evidence. The mechanism is strikingly different in the two rival explanations, since it pits the mechanism of natural selection against the mechanism of deliberate design. Natural selection is testable in principle and has been tested under laboratory conditions (Aids virus, coloration in guppies) and observed in nature (as in the case of industrial melanism). But acts of design are untestable in principle and will command little credibility in the scientific community because they require a leap of faith.

Blind Cave Insects

There are many examples in *The Origin of Species* in which Darwin uses some form of IBE, particularly IMF, in comparing his own theory of Natural Selection (NS) with the rival theory of Special Creationism (SC). The facts gathered from biology and geography may differ but the form of inference is the same in each case. As an illustration consider the remarks Darwin makes concerning blind insects in caves in North America and Europe.[5]

> It is difficult to imagine conditions of life more similar than deep limestone caverns under a nearly similar climate; so that on the common view of the blind animals having been separately created for the American and European caverns, close similarity in their organisation and affinities might have been expected; but, as Schiödte and others have remarked, this is not the case, and the cave-insects of the two continents are not more closely allied than might have been anticipated from the general resemblance of the other inhabitants of North America and Europe. On my view we must suppose that American animals, having ordinary powers of vision, slowly migrated by successive generations from the outer world into the deeper and deeper recesses of the Kentucky caves, as did European animals into the caves of Europe. We have some evidence of this gradation of habit; for, as Schiödte remarks, "animals not far remote from ordinary forms, prepare the transition from light to darkness. Next follow those that are constructed for twilight; and, last of all, those destined for total darkness."By the time that an animal had reached, after numberless generations, the deepest recesses, disuse will on this view have more or less perfectly obliterated its eyes, and natural selection will often have effected other changes, such as an increase in the length of the antennæ or palpi, as a compensation for blindness. ([7], p. 138, 1st edition online).

Darwin's argument in support of his theory of NS is comparative involving not only some facts about insect species but also a rival theory, SC; it also employs IMF and consilience. How are we to understand the theory of SC applied to this example? We are to suppose (a) that a Creator exists who wanted blind insects to occupy limestone caves with similar climates (along with other similar features) but in different parts of the world, such as America and Europe, and (b) the Creator has the power to bring about what he wants (so the insects do not evolve but are created). Granted SC would we expect the insects in similar caves around the world to be the same or different? SC cannot give a specific answer to this as we have no access to the mind, particularly the intentions, of the Creator. But the most obvious auxiliary assumption would be based on the principle of parsimony; just one basic design is used to create blind insects that "fit" well in similar cave environments around the world. Whether they are "best" fit or merely sufficient for survival can be left open.

We have no reason to make the less parsimonious assumption that the insects in similar but geographically widely separated caves and occupying similar niches would have two or more different basic designs. To suppose this would be to require that there be further matters to uncover concerning the intentions of the Creator which explain why there are different basic designs for blind insects living in similar caves in America and Europe. But there are no further matters to consider since, by the very nature of the supposition, we have no access to the intentions of the Creator. A further assumption that some gratuitous gesture of *joie de vivre* on the part of the Creator leads to different designs in similar caves around the world is

---

[5]This case study is also instructively discussed in Kitcher [9, pp. 45–46] as an example of Darwinian explanationism at work and its use as a critical tool for dealing with creationism.

of no help. Given the fact that living thing X exists in some environment (such as blind insects in caves), SC boils down to the claim that the Creator must have wanted X to be in that environment, must have had the power to realise his wants and the further auxiliary assumption that he used the one basic design rather than two or more.

In the above passage Darwin talks somewhat psychologistically about what we might expect or anticipate if we entertained SC (or NS). It is helpful to understand Darwin's talk of expectation in terms of probability, viz., the probability some hypothesis confers on some fact (that is, the likelihood of the hypothesis). It can be left open as to how the probability might be further understood (subjectively as a rational degree of belief, or something more objectivist).

Granted this, the claims based on the supposition of SC along with its extra auxiliary hypothesis can be expressed as the following likelihoods:

1. prob(same design for blind insects in similar caves in America and Europe, SC) is high;
2. prob(different design for blind insects in similar caves in America and Europe, SC) is low.

A second question can be posed for SC: though insects are similar in broad respects (e.g., common general features such as ability to absorb nutrients, etc.), in more specific respects would the blind insects in a cave be similar to, or different from, the sighted insects living outside the cave? Again, it is hard to know how to apply the theory of SC in this case since we are ignorant of what the Creator wants. But the idea of *separate* creation suggests that the design of each species of living thing is *independent* of the creation of another. Given information about the more specific features of one species nothing can be inferred about the more specific features of another species. Extending this to the blind insects in caves and the sighted insects outside the caves, that they are largely similar is as equally probable as their being dissimilar.

This yields a further auxiliary assumption of SC, that of the *independent, separate* creation each species:

3. prob(internal and external insects are similar, SC) = prob(internal and external insects are dissimilar, SC) = ½.

Alternatively if emphasis is put on the considerable environmental differences inside and outside the caves, one might expect that the Creator would not use the same basic design in making insects to fit such different environments; different basic designs would fit the different environments. Granted this, one would expect dissimilarity rather than similarity. Putting these two claims together we can say:

4. prob(blind insects in caves are similar to (some) insects external to caves, SC) ≤ ½.

What is denied here is that, given SC, the probability that the internal and external insects are similar is *greater* than the probability that they are dissimilar. That is, the probabilities of (3) and (4) cannot be greater than ½. They would be greater than ½ if one employed a different auxiliary hypothesis, viz., that the Creator is a "Tricky Creator" who makes it *appear as if* evolution occurred.[6] Granted this, the inside and outside insects would have strong similarities. But this is another gratuitous assumption about Creator intentions for which we could have no grounds (other than to save SC from refutation or to enable it to keep up with the more successful research programme of NS).

Now turn to the facts Darwin mentions about blind insects in caves around the world. These insects will have broad features in common, but beyond this they can differ markedly. The first fact is that the blind insects in similar caves in America and Europe do not strongly resemble one another; "they are not more closely allied than might have been anticipated from the general resemblance of the other inhabitants of North America and Europe." Unfortunately as (2) shows, SC assigns a low probability to this fact. But NS assigns a high probability, as Darwin indicates:

5. prob(different design for blind insects in similar caves in America and Europe, NS) is high.

Using IMF we can say, since SC has made a bad job of accommodating this fact about blind cave insects while NS does quite well, that NS is to be favoured over SC.

The second pair of facts is that the blind cave insects in American caves are similar to sighted insects in the surrounding external environment (and not those in Europe); and the blind insects in European caves are similar to those in their respective surrounding external environment (and not those in America). As (3) and (4) show, for SC this is either a matter of indifference or it is improbable (less than half). In contrast, on NS these facts are highly probable. Both Darwin and Schiödte tell us how this is so on the basis of NS. Sighted insects in the external environment with sunlight gradually moved into caves where there was very little or no light yet sufficient food to guarantee reproductive survival. In support of this, not only can the transitional forms be observed in some cases, but there is also an explanation of why some blind insects have much longer antennae than do their related forms living outside (to assist movement not guided by sight). So

6. prob(blind insects in caves are similar to (some) insects external to caves, NS) is high.

To sum up, Darwin considers two classes of facts: (F1) there is no strong similarity between blind cave insects in different parts of the world; (F2) there is strong similarity between insects internal to American caves and (some) insects external to these caves; and similarly for insects inside and outside caves in Europe. NS offers a good explanation of both classes of facts. In contrast SC can offer no good explanation and must accept each of these as unexplained brute classes of facts. Using IMF, we can infer in favour of NS and against SC.

---

[6]What is wrong with the "Tricky Designer" hypothesis? This point is not discussed here, but for a critique see Sober ([14], Chap. 2.6), 'The Problem of Predictive Equivalence'.

To complete the argument for NS providing a better explanation than SC, the definition of better explanation in section "Some Forms of Inference to the Best Explanation (IBE)" requires that the relative prior probabilities of NS and SC be taken into account. This can arise from the success of NS in accommodating the two distinct classes of facts, F1 and F2. NS consiliates, or unifies within the one theory, these two different classes of facts. In contrast for SC these two classes of facts are not made consilient; they remain independent of one another. Such consilience adds to the confirmational support that NS gets over SC. So not only does NS make these two classes of facts more likely than SC does, it also consiliates these two classes of facts while SC does not, thereby gathering extra conformational support. Taking these two claims together concerning likelihoods and consilience, we can argue that NS gives a good explanation of the facts while SC gives no explanation of them.

Many other such examples[7] of the use of IMF can be found in *The Origin of Species*. In the light of these one can clearly understand why Darwin describes his book as one long argument; it is an ever expanding set of classes of facts from biology and geography made consilient by NS which also serve as a grand conjunctive premise for an IMF inference to the conclusion that NS is to be favoured over SC.

Conclusion

Probability considerations may shift the balance of credibility of theories; the evidence bestows credibility on one theory while at the same time discrediting its rivals. In the *Origin of Species*, Darwin uses comparative probability arguments, like IMF, rather than the method of falsifiability. The Darwinian inferences take the following form: The evolutionary theory confers more probability on the evidence than the rival creation theory. Hence, on Darwin's account the creation theory should be eliminated because it fails to provide a mechanism, which renders the evidence probable.

We should finally consider how IMF is related to falsificationism. Critics of Popper's falsificationism point out that the hypothetico-deductive scheme does not lead to the exclusion of rival theories, since one theory at a time faces the evidence. But this leaves open the possibility that there are alternative theories, which may either be equally well corroborated or even be in better agreement with experimental evidence than the theory under test. By contrast, IMF achieves a reduction of a set of rival accounts by letting principles and evidence either lower the probability of the rival theory or eliminate unsuccessful theories altogether ([20] pp. 1–6; [21], pp. 11–13). For instance, the simple fact of 'being at location A at time t' when

---

[7]As a sample (a search will reveal more) Darwin compares favourably his theory of Natural Selection with respect to Special Creationism in ([7], first edition, pages 159–167, 185–186, 194, 377–379 and 434–435).

a crime happened at location B at time t eliminates a whole raft of potential suspects. In the same way all theories based on the notion of Intelligent Design or creation are eliminated by Darwinian inferences. Popper was aware of this consequence of the falsification principle. Only when a limited number of theories is available does the critical method lead to the elimination of all unfit competitors, by the method of falsification. In the normal case, the number of competitors is large and the critical method cannot drive the elimination method to a point where only one 'true' competitor survives ([22], p. 16, cf. pp. 264–265; [23], pp. 107–108).

There exists thus a limit relation between IMF and Popper's hypothetic-deductive model. When the set of alternative or rival theories can be regarded as very restricted, for instance on account of criteria like 'simplicity' and 'coherence', falsificationism may be regarded as a limit of the method of IMF. In the limit of a paucity of choice, for whatever reason, IMF goes over into falsificationism. But such situations are rare in the history of science, as Popper's model of the growth of scientific knowledge would suggest. In the case of Darwinism there were, and there continue to be, a number of rivals and alternatives [24].

Acknowledgement Parts of sections "Some Forms of Inference to the Best Explanation (IBE)" and "Blind Cave Insects" are based on material in a paper by R. Nola 'Darwin's Arguments in Favour of Natural Selection and Against Special Creationism' to appear in Science & Education' also published by Springer.

References

1. Ghiselin, M.T.: The Triumph of the Darwinian Method. University of California Press, Berkeley/Los Angeles (1969)
2. Ruse, M.: The Darwinian Revolution. The University of Chicago Press, Chicago/London (1999)
3. Sober, E.: Evidence and Evolution: The Logic Behind the Science. Cambridge University Press, Cambridge (2008)
4. Lakatos, I.: The Methodology of Scientific Research Programmes: Philosophical Papers, vol. I. Cambridge University Press, Cambridge (1978)
5. Ladyman, J.: Understanding Philosophy of Science. Routledge, London/New York (2002)
6. Gillies, A.: An Objective Theory of Probability. Methuen, London (1973)
7. Darwin, C.: The Origin of Species. The complete work of Charles Darwin (1859-72) Online for all six editions: http://darwin-online.org.uk/contents.html#origin
8. Bacon, F.: In: Lisa Jardine/Michael Silverstone (ed), Novum Organum (1620). Cambridge University Press, Cambridge (2000)
9. Kitcher, P.: Living with Darwin: Evolution, Design and the Future of Faith. Oxford University Press, New York (2007)
10. Lewens, T.: Darwin. Routledge, London (2007)
11. Niinuluoto, I.: Abduction and truthlikeness. In: Festa, R., Aliseda, A., Peijnenburg, J. (eds.) Confirmation, Empirical Progress and Truth Approximation, Poznan Studies in the Philosophy of Science and the Humanities, vol. 89, pp. 255–275 (2005)
12. Peirce, C.S.: In: Buchler, J. (ed.) Philosophical Writings of Peirce. Dover, New York (1955)
13. Musgrave, A.: Essays on Realism and Rationalism. Rodopi, Amsterdam (1999)
14. Sober, E.: Philosophy of Biology. Boulder Co., Westview (1993)

15. Glass, D.: Coherence measure and inference to the best explanation. Synthese **157**, 275–296 (2007)
16. Van Fraassen, B.: The Scientific Image. Clarendon, Oxford (1980)
17. Salmon, W.C.: Reality and Rationality. Oxford University Press, New York/Oxford (2005)
18. Butts, R. (ed.): William Whewell Theory of Scientific Method. Hackett Publishing Co, Cambridge (1989)
19. McGrew, T.: Confirmation, heuristics, and explanatory reasoning. Br. J. Philos. Sci. **54**(4), 553–567 (2003)
20. Norton, J.D.: Science and certainty. Synthese **99**, 3–22 (1994)
21. Norton, J.D.: Eliminative induction as a method of discovery: how Einstein discovered general relativity. In: Leplin, J. (ed.) The Creation of Ideas in Physics, pp. 29–69. Kluwer, Dordrecht (1995)
22. Popper, K.: The Logic of Scientific Discovery. Hutchinson, London (1959). English translation of *Die Logik der Forschung* 1934
23. Popper, K.: Objective Knowledge: An Evolutionary Approach. Clarendon, Oxford (1972)
24. Weinert, F.: The role of probability arguments in the history of science. Stud. Hist. Philos. Sci. **41**, 95–104 (2010)

Breaking the Bonds of Biology – Natural Selection in Nelson and Winter's Evolutionary Economics

Eugene Earnshaw-Whyte

Introduction

Nelson and Winter's *An evolutionary theory of economic change* purports to offer a new foundation for economic theorising; the title indicates where they draw their inspiration for this bold enterprise. Their work represents an early and foundational contribution to the now thriving subfield of evolutionary economics. Whatever the virtues or flaws of Nelson and Winter's work from the perspective of economics, I am not an economist and my interest is philosophical: I am concerned with evolution by natural selection ('ENS'), considered generally. I take it that if we wish to understand what natural selection is in the abstract, models of evolution that diverge from the familiar may be instructive. Nelson and Winter's models diverge from the familiar in significant ways, although not so far as to render their similarity with models of biological evolution opaque. They therefore provide an interesting subject of philosophical analysis – they present sophisticated mathematical models, self-consciously evolutionary in character, which allow us to examine and perhaps refine the traditional assumptions made by philosophers of biology about the nature of evolution by natural selection, as well as understanding how evolutionary models can be constructed in the context non-biological disciplines.

Any genuine attempt to employ Darwinian principles to explanations outside the domain of biology has considerable interest from a philosophical standpoint, precisely because of the overwhelmingly biological character of most models of natural selection. It is widely accepted that the explanatory framework of Darwinian evolution is general enough that it could be, in principle, applicable in

E. Earnshaw-Whyte (✉)
PhD Candidate, Institute for the History and Philosophy of Science and Technology, University of Toronto, Ontario, ON, Canada
e-mail: malefax@rogers.com

M. Brinkworth and F. Weinert (eds.), *Evolution 2.0*, The Frontiers Collection,
DOI 10.1007/978-3-642-20496-8_10, © Springer-Verlag Berlin Heidelberg 2012

many domains[1]; it is also undeniable that its greatest success and widest adoption is in the field to which it was first applied. However, there are serious unresolved issues concerning the nature of selection and its relation to heredity, to novelty, and to evolution generally. As much as many authors have tried to develop a more general framework for evolution by natural selection [4, 7], the theoretical underpinnings of ENS were developed by philosophers of biology in order to resolve biological questions. Insofar as a philosopher might wish to understand evolution by natural selection in its more general form, examining evolutionary explanations that diverge as widely as possible from the familiar form of population biology should be instructive.

I begin by developing a framework of analysis in terms of the interaction of evolutionary mechanisms, which resolves some difficulties that otherwise arise in interpreting evolutionary change. I then apply this framework to Nelson and Winter, discussing and interpreting one of the models they introduce in their 1982 book in considerable detail. This analysis confirms and illustrates the analysis of evolutionary change in terms of discrete mechanisms. Based on the analysis, I conclude that neither drift nor selection are distinctive mechanisms of change, and there is no clear necessary demarcation between novelty-producing mechanisms and selection mechanisms. In particular, the 'search' mechanism in Nelson and Winter can, under appropriate circumstances, serve as a selective mechanism in addition to providing novelty to the system.

Evolution How?

In presenting their model as evolutionary, Nelson and Winter do not merely intend to suggest that economic change involves the unfolding of some regular pattern. That sense of evolution has a long history in the social sciences: Lewontin and Fracchia [8] refer to it as 'transformational' evolution: it is the evolution of Lamarck and Spencer. That kind of evolution is rooted in the internal development of individuals; the individual changes over time either in accordance with some internal organising principle, or in response to the influence of the environment. As the individual changes, so does change in the population reflect the predictable development of the individuals that comprise it. Indeed, in a transformational concept of evolution, it is the individual that evolves: the transformation of the individual might be explained by the characteristics of the parts that comprise it, but one treats the evolving unit as a whole engaged in a transformational developmental

[1]Sober has stated this repeatedly [1, 2]; Dawkins [3] popularised the notion with his concept of 'memes', it is defended at length in Aldrich et al. [4]; see also Taylor in this volume. Criticisms of evolutionary ideas outside biology tend to be of the specifics, not of the principle, although see Foster [5] and Witt [6] for arguments against the applicability of 'biological analogies' to economics.

process, whether it is a species, a social group,[2] a 'race', or a galaxy. Many 'evolutionary' models in the social sciences are evolutionary in the transformational sense, lacking any real connection to Darwinian modes of explanation.[3]

Aside from the 'transformational' concept of evolution, the term is also employed in another non-Darwinian sense. This is of slow and incremental change; change that may involve intelligent and deliberate decisions, but which unfolds according to patterns of a scope beyond the ken of those deciders. The change has a particular directional pattern associated with it, a pattern that demands explanation. The large-scale pattern, if it is an evolutionary one, passes through certain stages in a predictable order. Historically this concept is associated with the idea of progress – social progress, moral progress, scientific progress – for to the nineteenth century eye, such progress was an observed fact that required explanation. This 'directional' evolution can be (but need not be) explained via 'transformational' evolution: the slow pattern observable in the historical record results from the unfolding of individual developments. This was how Lamarck, for example, explained the pattern of progress he observed in the fossil record: individual species gradually transform themselves according to internal principles of improvement (including adaptation to their environment, passed on to their offspring). Similarly, one can explain the increase in the efficiency of an industry over time as the result of the internally driven development of the firms that comprise it. Following Darwin, however, this is not the approach that Nelson and Winter take.

Evolution by Natural Selection

The Darwinian sense of evolution, dominant in biology and employed by Nelson and Winter, concerns the change over time not of an individual, but of a population, and explains the dynamics of this population in a characteristic way. Sober [1] characterises this as a 'theory of forces' and compares it to Newtonian mechanics. The idea is that the population is influenced by distinct factors that jointly generate evolutionary change. The evolution itself consists of the change in the frequency of particular traits[4] in the population. The most characteristically Darwinian of the factors that generate change in the population is 'natural selection', a process that biases the population towards those traits which are advantageous for survival and reproduction. The process is of variants in struggle, expanding or declining within the population at the expense of alternative variants. Evolution by natural selection (ENS) explains, among other things, how a system can move in particular predictable direction over time despite none of the parts having any intention of doing that.

[2]Perhaps engaged in evolving from 'savagery' to 'barbarism'.

[3]Sanderson illustrates this in *Evolutionism and its Critics* [9].

[4]Evolution is often defined as change in *gene* frequencies (see [10, 11] for instance), but this is obviously unsatisfactory outside of biological contexts.

Its explanatory tool in this undertaking is competitive advantage: in the environment in question, some variants are more successful than others, and this success feeds back into the subsequent prevalence of the variants in question.

This competitive advantage consists in some stable[5] tendency of the type to grow, reproduce, survive, convert, or otherwise expand their representation in the population as a whole; this being usually referred to as 'superior fitness'.[6] For Darwin, competitive advantage was conceived as belonging to the individuals in virtue of their particular traits and the state of the environment. For example, if fleet prey (deer) become more abundant, the swiftest wolves have an advantage, swift wolves will tend to be preserved and slow ones destroyed, and this will affect the species as if wolves were being bred for swiftness.[7] The trait (swiftness) leads to the individual surviving, which over time causes the population as a whole to have more swift wolves in it. Individual wolves are selected in the sense of being preserved, but what actually spreads in the population is the generic trait of swiftness. For this basic sort of natural selection explanation, the explanandum is the spread of the type in the population, and the explanans is the propensity of individuals of that type to succeed better than individuals of other types. While there may be other, similar ENS explanatory modes,[8] this kind of explanation captures nicely both how Darwin seems to understand ENS, and what goes on in Nelson and Winter's models, so I will suppose that it at least suffices for an explanation to be an ENS one.

Insofar as we wish to explain directional evolution by appealing to ENS, all that is required is for change in that direction to be correlated with the expansion of the successful types. This correlation may, but need not, be rooted in a direct causal link between the trait of interest and success. So, for example, if having bigger antlers is correlated with success in male deer, this can explain an increase of the size of deer antlers over time – regardless of whether the antlers are directly causing the success or are by-products of some other success-enhancing trait. So, in this mode, one might explain the spread of mechanisation during the nineteenth century by its correlation with superior profitability in much the same way as one explains the subsequent spread of black moths in the resulting sooty environment. In both cases, the trait of interest is correlated with success; this explains its increase in prevalence over time.

The explanation of *why* the trait of interest is correlated with success is a separate question, and it is unlikely to be answered by appealing to evolution. Rather, answering it requires some sort of plausible story about what makes for success

[5]In the environment of interest: competitive advantage is always environment relative.

[6]Although 'superior fitness' is often intended to refer specifically to *reproductive* success, hence my preference for the broader term, 'success'.

[7]Darwin, 1859 [12, p. 90].

[8]See Ariew and Matthen [13] (2002), Godfrey-Smith[14], or Hull, Langman and Glenn [7], for various recent attempts to characterise ENS generally. My approach here is to give a sufficient condition for ENS explanation without trying to give an exhaustive definition.

in the environment, and why the individuals with a tendency to be successful also tend to have (more of) the relevant trait. This in turn demands a domain specific analysis, which will appeal to the particular environment, the different traits present in the population, the nature of the interactions between individuals and the environment, and so forth. So although one can explain the evolution (in the sense of directional evolution) of a population by appealing to evolution (in the sense of ENS), part of a more thorough understanding of the nature of the change will still advert to a non-evolutionary analysis of the causal relationships of the system in question.[9]

Novelty and Explanation

While evolution via natural selection may suffice for certain explanatory purposes, much of the explanatory power we associate with Darwin requires that selection be supplemented with a source of novelty.[10] This does not refer to *variety*, without which natural selection is quite impossible. Variety is distinct from novelty, in the sense that the diversity in a population is distinct from the *source* of that diversity. Natural selection strictly speaking requires variety but not novelty; novelty, however, was an essential element in Darwin's own evolutionary explanations, and offers distinct and desirable possibilities to other theorists – in particular, it has been widely recognised as a key element in the explanation of technological change and innovation, a central concern of evolutionary economics [16, 17]. For this reason, to avoid confusion with other sorts of evolutionary thinking, I will refer to evolutionary models involving the combination of ENS and novelty as 'Darwinian'.

Each of the major explanatory triumphs of Darwin's *Origin of Species* relies crucially on both a source of novelty and the action of natural selection contributing to the overall change in the population. The production of complex adaptations, for example, requires both that new variants continually emerge, and that such variants as are improvements on the complex organ spread in the population at the expense of the less adapted version. Over time, the continual interaction of novelty and selection produces the organic marvel of the eye. Similarly with his explanation of speciation: novelty and selection combine to produce variants or 'races', which over time can become well-marked species through the wandering interaction of novelty and selection.

The source of novelty in Darwinian evolutionary explanation has a somewhat paradoxical character, in that it is an essential part of the explanation, but its details are outside the explanation's scope. Darwin himself observed and argued that novelty arises spontaneously in living population; but he knew nothing of its

[9]This is emphasized in Sober [2].

[10]Hull, Langman, and Glenn [7] particularly emphasize the importance of a source of novelty in their general account of evolution, in contrast to versions such as Godfrey Smith [14] and Lewontin [15] that do not highlight a distinct role for novelty as compared with mere variety.

origins. It was enough for him that a homogenous population would not stay purely homogenous from generation to generation: nature was an inexhaustible font of novelty. This is analogous to the fact that success in the environment is an element of an ENS explanation that may not be itself explained: understanding why a trait conduces to success is not required in appealing to the influence of the trait on the evolution of the population.

Mechanism

We have discussed how Darwinian evolution can explain directional evolution, or the spread and decline of different variants. This can be conceived in a loose, verbal way, or in a precise, quantitative way. Darwin himself explained in the verbal mode; the point was to persuade his audience of what novelty and selection could achieve in tandem. But since his time, evolutionary explanations have been mathematised. This was a key component of the neo-Darwinian Synthesis, founded on the development of techniques for dealing with statistical population dynamics by authors such as R.A. Fisher and J.B.S. Haldane.

Broadly, a quantitative approach to evolutionary explanation requires that one specify the sources of change in a population. So, for example, one might specify the different alleles, and give a fitness value for each genotype; the fitness value, if it is fully specified, gives the probability distribution of reproductive success. Given a population so specified, we have a Markov process: we can calculate the probability of any subsequent possible state. A mathematical formula so specified as to constitute a Markov process will be henceforth referred to as a 'mechanism'; the term is loosely analogous to the 'forces' conceived by Sober [1]. The total model of change may be a composite of Markov processes each sufficient to give a probability distribution on its own. So in addition to the fitness mechanism, we could introduce a mechanism of mutation, which specifies the probabilities that individuals possessing one allele might spontaneously change to possessing a different one. If this is fully specified, we will again have a Markov process, this time deriving from the probabilities associated with both of these distinct mechanisms.[11] And there is no particular limit to how many mechanisms of change might be introduced. So we might, for example, have different mechanisms that model mutation due to transcription error, due to cosmic radiation, and due to environmental toxins. Jointly, they specify a Markov process; individually, they are causes of evolution.

It is often presumed that natural selection is bound to be one distinct mechanism among those that jointly constitute the evolutionary process,[12] but it may already

[11]The mechanisms being 'distinct' insofar as each individually would suffice for a Markov process.

[12]Sober's discussion [1] suggests this, and it is assumed in Stephens' defence of Sober's force model [18].

be clear that this cannot be the case. Firstly, just as mutation might be most perspicuously modelled via several distinct mechanisms, corresponding to distinctive causes of alleles changing into different alleles, so might the reproductive output of the population be most perspicuously modelled using several distinct mechanisms. For example, some species both reproduce sexually and via parthenogenesis. Given the different hereditary patterns involved, it would likely be wise to model each type of reproductive success separately. An allele could be advantageous with regard to sexual reproduction but disadvantageous for parthenogenesis. In this case, selection is operating in one direction via the one mechanism, oppositely via the other, and can be directly identified with neither.

For any mechanism in the model, given the environment, the operation of the mechanism will either tend to change the relative prevalence of types in the population towards some types at the expense of others, or not. This tendency, if it exists, might or might not be realised (if change is probabilistic, for example, the population might change in the less-likely direction), and it might either be reinforced or counteracted by the operation of other mechanisms. But the directional bias in the mechanism exists independently of the change that actually occurs, just as in the traditional understanding there may be selection pressure even if the population drifts in the opposite direction. The contrast is that the traditional understanding identifies selection as one specific evolutionary force out of many, whereas on the account developed herein, any mechanism is selective in a given environment just insofar as its operation tends to change the relative prevalence of alternative variants.

In the context of a population model generally, novelty refers to any mechanism that can introduce types into the population that did not previously exist – which is what 'Darwinian' evolutionary explanation requires. Insofar as selection operates by skewing the population toward *more* of the fitter and *less* of the less fit, it does not produce novelty. One therefore initially expects separate mechanisms of novelty and mechanisms of selection, insofar as we expect a selection mechanism to act on pre-existing variety in the population, whereas a mechanism of novelty produces variety out of nowhere, so to speak. Further analysis, however, suggests that mechanisms can fulfill both these roles simultaneously.

A novelty producing mechanism is just one where the operation of the mechanism can introduce types into the population that did not previously exist – ideally one that can produce vastly more possibilities than exist in the population, or than can ever be actualised. By contrast, a mechanism is selective just insofar as its operation tends to favour one variant at the expense of alternative variants. And there is no reason a single mechanism can not meet both of these desiderata. For example, if one allele of a gene has a strong tendency to mutate into its rival, but the opposite allele is much more stable, the process of mutation will tend to bias the population in favour of the stable gene. It is clear that the gradual and predictable spread of the stable allele at the expense of the unstable one is explained due to its competitive advantage in the environment, as mediated by the mechanism of mutation. This mutational mechanism may also allow for the possibility of entirely new alleles arising as well, making it simultaneously a mechanism of novelty.

Insofar as by appealing to 'natural selection' we are engaged in explaining the directional tendencies of change in the population by appeal to the competitive advantages of particular traits, then by appealing to the directional bias of this mutation mechanism, we are engaged in natural selection explanation. Another way of putting this is that natural selection occurs whenever there is variation in 'trait fitness' – the all-things considered tendency of a type to increase at the expense of its rivals [19]. But variation in trait fitness does not always trace back to variation in survivability or fecundity. The stability of one variant of a gene compared with another can underwrite a difference in trait fitness that causes evolution by natural selection.

It might be possible to isolate only such mechanisms as operate via something intuitively like 'reproduction' or 'survival' and restrict the use of the term 'selection' to the biasing effects of such mechanisms. If this approach is preferred, we would require a different term to discuss the total biasing effects of the mechanisms operating in the model: it would be appeal to this broader term that would explain the predictable patterns of change in the traits of the model. I will persist in using the term 'selection' in the broader sense: among other things, this avoids the difficulty of specifying what 'reproduction' amounts to in non-biological contexts.

Interpreting Nelson and Winter

Nelson and Winter's evolutionary economics represents an attempt to challenge what they characterise as the dominant orthodoxy in economics: the neo-classical school. They are critical of several of the core methodological assumptions of neo-classical economics, in particular that economic agents maximise profits via choice from a well defined set of alternatives. They consider the case of innovation to be particularly problematic for neo-classical economics as innovation upsets the equilibrium that neo-classical modes of analysis assume to prevail, and innovative possibilities are not known and understood in advance.

It is partly this desire to model corporate innovation that led Nelson and Winter to develop their evolutionary economics.[13] In developing these views, they were working in the long tradition of evolutionary theorising in economics and the social sciences more broadly, in particular the economists Shumpeter and Hayek [20]. Nelson and Winter's program is particularly distinctive with regard to the extent to which they explicitly take biology as a model for their enterprise, commenting on what they take to be analogues between elements of their theory and the biological theory of natural selection. This close and conscious analogy is part of what makes their book of particular interest from the point of view of philosophical analysis.

[13]Nelson and Winter [10, p. 129]; see also [20].

They both present broad methodological principles that could guide further research, and develop specific models to investigate particular issues – in particular, they are interested in showing that an evolutionary economic model can do just as well as neo-classical economics in modeling historical episodes of economic change. This is demonstrated in Chap. 9 of *An evolutionary theory of economic change* [21] by taking the economic conditions at a particular time and, employing plausible assumptions, showing that repeated simulations generate change over time quantitatively similar to the actual historical changes.

As this model is the most detailed of those presented in *An evolutionary theory of economic change*, it will be the focus of the analysis. The model is intended by Nelson and Winter to test the ability of their approach to "predict and illuminate" [21, p. 206] economic patterns of growth to a comparable degree as neoclassical theory. In practice, the idea is to take historical data and build a plausible model that, given starting values similar to those characterising the industry in a particular year historically, predicts economic growth which closely resembles the actual history. The question is "whether a behavioural-evolutionary model of the economic growth process... is capable of generating (and hence of explaining) macro time series data of roughly the sort actually observed" [21, p. 220]. This is fairly ambitious by biological standards: population genetics doesn't generally try to recreate historical episodes of adaptation.[14]

The model, discussed in Chap. 9 of their book, posits an industry composed of a number of competing firms. Each firm is characterised by a production technique and a capital stock. The production technique sets the amount of labour and capital required to produce a given output of GNP – these are the 'input coefficients' that characterise the productive activity of the firm. The capital stock sets the total productive capacity of the firm, and measures its 'size' in the industry: a firm with no capital stock is not actively in business, since it produces nothing and employs no-one, but it is modelled as engaged in research, looking for a technique that could be employed profitably in the prevailing economic environment.[15] The total state of the industry – that is, a full description of the 'population' – is given simply by the capital stock and production technique of each firm.

In this model there are two operative mechanisms of change. Any change – and therefore anything we might be tempted to call 'evolution' – must occur via the operation of either or both of these mechanisms. One is the mechanism whereby firms increase or decrease their capital stock based on their profitability – which I shall refer to as the 'capital change' mechanism. This change in firm capital stock is compounded of two factors: a fixed depreciation rate per unit of capital stock, and

[14]Compare Fisher' *Genetical Theory of Natural Selection*, [11] for example, which is almost exclusively concerned with providing proofs-in-principle that natural selection could have certain sorts of effects.

[15]If such a firm does find a technique, it may re-enter the market with a small stock of capital.

firm investment, which is equal to firm profits[16] (and can be negative). The consequence of the capital change mechanism is that a firm needs to be slightly profitable in order to maintain itself at a given size, and profits or losses feed back directly into the size of the firm.

A key factor influencing this mechanism is the price of labour. This is set by an equation, which takes total labour used in a time period as its main variable, although time could be also included (allowing, for example, the situation to be modeled where labour tends to become cheaper over time).[17] This equation can be understood as modelling the influence of an aspect of the ‘environment’ on the industry. Since costs directly subtract from the growth of the firms, here the environmental influence incorporated into the model functions purely as a negative pressure on growth.

The production technique of a firm is, as mentioned previously, characterised by just two coefficients: labour input, and capital input. These, together with the price of labour discussed above, are the only variables that influence a firm's profitability. The capital input coefficient together with the capital stock of the firm determines the output of the firm, which directly determines the firm's gross revenue, from which profits are determined. The labour coefficient sets a cost based on the firm's output and the labour cost (which is a function of the industry's output as a whole). Both of the production technique coefficients therefore directly, and separately, influence the profitability of the firm, but the relative importance of each is determined by the cost of labour: if labour is relatively cheap then the productivity of the capital stock of the firm will have a larger influence on the overall profitability of the firm, but if it is expensive then the productivity of labour given by the technique will tend to have the dominant effect on the firm's growth. This means that a technique may be superior to its rival under some market conditions, but inferior in others. Production techniques with better (lower) values for both capital and labour input are, of course, strictly superior to their comprehensively less-efficient rivals.

This capital change mechanism is naturally viewed as a process of selection.[18] The average rate of growth or decline of a firm is a function of its profitability per unit capital. The firms best suited to the given environment[19] therefore increase their overall market share at the expense of their less efficient rivals. Furthermore,

---

[16]Profits are given by firm revenues (equal to firm output, determined by its production technique and capital stock), minus its labor costs, minus its required dividend (a function of its capital stock; this parameter is varied in the simulations run by Nelson and Winter).

[17]Nelson and Winter leave their demand function as constant over time: I make use of the possibility of an influence of time on labour costs subsequently.

[18]In situations where there is variety among the existing techniques, at least. This highlights the fact that any mechanism is only ‘selective’ under certain circumstances, further undermining the appropriateness of identifying selection with any particular mechanism.

[19]Either by being strictly superior to their rivals under all circumstances, or by having a superiority with regard to one of the input coefficients that outweighs inferiority in the other under the prevailing cost of labour.

any firm's profitability (and therefore its growth) is directly and negatively impacted by the success of other firms: the growth of other firms increases the amount of labour employed and therefore the cost of labour. Changes in the relative size of firms, in this model, are almost exclusively attributable to differences in profitability per unit capital,[20] and these differences are directly a function of the differences between the input coefficients of the production technique of each firm. We can explain the expansion in the market share of a particular type of production technique in just the same way Darwin explains wolves becoming swifter when deer become more prevalent in the environment[21]: the individuals in the population vary with regard to a trait that causally impacts success; success (whether bringing down deer or generating large revenues) causes the trait of the successful individuals to become more prevalent in the population; the trait that conduces most to individual success therefore becomes widespread.

Of course, swift wolves become more prevalent in the population in terms of numbers, whereas firms become prevalent in the industry in terms of size. This is remarkable precisely because the explanation is the same in both cases. The pattern is realised through very different causal mechanisms, but this is also the case in biology: the causal factors and mechanisms whereby individual wolves survive and reproduce are very different from those relevant to the spread of aphids, let alone bacteria.

Search

The other mechanism is 'search', which is responsible for the firms replacing old techniques with new ones. Only unprofitable firms engage in search in this model: Nelson and Winter present this as a 'conservative' assumption, suggesting that it is interesting to show that innovation can importantly influence industry change even if individual firms are not especially motivated to engage in it. Alternate search rules – including an 'unbiased' search rule where all firms are equally likely to engage in search – are equally possible. In any case, since the 'capital change' mechanism simply acts on whatever variety is present in the industry without regard to how the variety came about, it makes no difference to the 'capital change' selection pressures what bias was involved in the operation of the mechanism of novelty.

Search is modeled as either involving new research ('local search'), or as the imitation of the production technique of a rival firm. A firm engaging in research

[20]The attrition of capital over time is a probabilistic process, so a firm can randomly shrink more or less than expected; this effect is more likely to be significant insofar as the firm is relatively small. Such effects are analogous to drift in biology, but are relatively small because most of the mechanism is deterministic.

[21]Darwin, 1859[12, p. 90].

has a chance of developing a new production routine: the characteristics[22] of this routine are determined by a draw from the possibility space of production techniques, biased so as to cluster around production techniques similar to the firm's current technique. A firm imitating others will tend to imitate the firms responsible for producing most of the industry output. In either case, the firm will test the prospective technique to determine if it will be profitable: this test is subject to a margin of error, meaning that a firm may adopt a new technique that is actually less efficient than the old. This aspect of the model introduces another aspect of 'bias', in the sense that advantageous techniques are more likely to be produced than disadvantageous ones. This means that the overall tendency in the model towards the dominance of increasingly efficient production techniques is not solely, nor necessarily even mostly, due to the tendency of relatively efficient companies to grow and inefficient companies to shrink. The tendency of unprofitable firms to engage in research, and for research to result in the adoption of more-profitable techniques, is sufficient to produce efficiency growth.

Nelson and Winter explicitly discuss the need for a search mechanism analogous to mutation in biology in order to "fill in the ranks of behaviour patterns decimated by competitive struggles... or to make possible the appearance of entirely new patterns" [21, p. 142]. Innovation, they state, is the analogue of mutation in their models, and in the model in question, 'search' brings about innovation. Nelson and Winter emphasize that "the set of potential routines that can be reached by search becomes a major analytic concern" [21, p. 143]. It is concerning precisely because of a difficulty that motivated Nelson and Winter to undertake their program: the difficulty of characterising in advance how innovation might proceed. For the particular model in question, the possibility space of techniques was generated from the historical data, chosen out of several alternative sets so as to be relatively free of 'holes' with regard to values. In cases where historical data is not available (such as when analysing the prospects of a contemporary industry), the problem of how to produce a plausible mechanism of innovation becomes acute. Biological models may face an even more severe difficulty in trying to characterise the operation of mutation quantitatively since it is so hard to analyse the possibility space of organism fitness, let alone connect that back to the set of possible heritable mutations and a plausible probability function thereof.

Search as Selection

These two mechanisms are each individually sufficient to generate an evolutionary process of change in an industry. Without the search mechanism, firms would grow or shrink depending on their relative profitability, with the efficiency of firms possibly changing as the total output (and therefore the labour costs) of the industry

[22]That is the labor input coefficient and the capital input coefficient.

changes. The change over time of the industry would be determined purely by the relative efficiencies of the unchanging techniques employed by each firm. Unless labour costs are set to decrease over time, the growth of the firms would eventually stabilise at a size where the costs of labour would just allow for tiny profits sufficient to counterbalance the depreciation of the capital stock.

If the search mechanism was the only mechanism of change, the size of the firms would remain constant over time. However, firms would still generate revenues. Any unprofitable firms would engage in search, and would alter their search techniques if they found techniques they took to be more effective. If labour costs remained constant or decreased as a function of time, change would soon cease as every firm would become profitable, making the behaviour of the model relatively uninteresting. If labour costs increased over time, however, any given production technique would eventually become unprofitable, forcing a search process for a more effective technique.

In this case, change in the industry is a matter of gradual innovation, where profitable techniques persist in the industry (and may be imitated by others) until the steady increase in labour costs render them obsolete. The capital stock of the firm is effectively irrelevant,[23] as there are no economies of scale in the model. The behaviour of the model is determined by the parameters of the search technique (the extent to which search tends to find 'nearby' techniques, for example, is set by a parameter which Nelson and Winter vary in different simulation runs), together with the environmental factor of labour costs. The effect of the production technique coefficients is to set, one might say, the survivability of the technique: techniques above the threshold of profitability will maintain their representation in the population of firms or increase, whereas although unprofitable techniques might survive (or even increase) for a time, they are liable to disappear at any point.

By analogy with biology, Nelson and Winter explicitly compare the capital change mechanism with selection, and the search mechanism with mutation. However, the search mechanism in Nelson and Winter plays a very central and active role in driving the increase of profitable techniques and the decline of less suitable alternatives. In a regime where search is the only mechanism, firms do not compete directly: their profits and their search patterns are isolated from those of other firms. As we have seen, however, the traits of the firms do actually influence their prevalence in the environment: profitable techniques persist and may be imitated while unprofitable techniques tend to vanish. Under certain circumstances, indeed, the behaviour of the model can be explained quite fruitfully in terms of selection even if 'search' is the only mechanism.

[23]There is a small effect, in that larger firms are more likely to be imitated than small firms, as firms tend to be imitated in proportion to their share of industry output. This will effectively be random, in the sense that firm size is set by the initial conditions and is unrelated to the historical success of the firm in terms of profitability. Firm size therefore would factor in the model only as an initially set 'imitation bias': one could set the initial size of the firms to be equal and remove this effect entirely.

Consider the situation where labour costs are a function of time, but one that fluctuates within certain bounds rather than being a simple increasing or decreasing function. Assume also that rather than engaging in search only when unprofitable, firms engage in search when their profits per unit capital are below the industry average. And assume that firms are relatively numerous, and that search for new techniques is relatively unlikely to find a new technique. Under these circumstances, below-average techniques will have a tendency to decline, whereas superior techniques will persist and possibly spread. However, environmental conditions (the cost of labour) may fluctuate, which can cause previously inferior techniques to become superior. Over time, we will tend to see the overall efficiency of the industry increase, as new techniques are adopted and spread through imitation. One could see an oscillation back and forth between two techniques, one with superior labour productivity, one with superior capital productivity, if labour costs repeatedly cross the critical threshold that determines which technique is the more profitable. As over time a technique approaches obsolescence,[24] gradually fewer firms will employ it as they either imitate techniques with a larger share of industry production, or else manage to develop a new and superior technique.

It should be clear that we can explain patterns of change in this model using the most basic sort of natural selection explanation: the traits of an individual determine its success in the environment, which influences the subsequent prevalence of those traits in the environment vis-à-vis its competitors. Success is again profitability, but here profitability influences not the growth of firms, but rather the spread of techniques. Making the decision to innovate governed by average industry profitability seems to be the key change that makes the scenario amenable to explanation by natural selection. This transforms industry situation into one wherein the firms influence one another, instead of developing in isolation. Indeed, the original situation is reminiscent of the 'transformational' sense of evolution discussed earlier, in that the development of each individual firm proceeds in near total isolation from outside influence. The alternative situation is one where we can understand the change in the industry as the result of selective pressures, precisely because the individual changes reflect interaction and competition between the members of the population.

This is despite the fact that the mechanism of change is still the same as that which provides the source of novelty in the models Nelson and Winter actually explore. The only alteration is in some of the parameters of the model (and, of course, the suspension of the 'capital change' mechanism). This illustrates is that it depends on the circumstances whether a mechanism causes selection, and that it is futile to try to characterise a mechanism in advance as 'the' mechanism of selection. With regard to the parameters actually employed by Nelson and Winter, both mechanisms operate so as to bias the population towards some techniques and away from others based on the success of the techniques. While the capital change mechanism operates in a manner more obviously analogous to 'fitness' in biology,

[24]Meaning an increasing proportion of the population employs strictly superior techniques.

the mechanism of novelty also acts so as to bias the population towards profitable techniques and away from unprofitable ones. Tweaking the parameters of the model actually only makes this process more obvious: it still operates to a limited extent in the regimes considered explicitly by Nelson and Winter, because of the tendency of only unprofitable firms to search, and because of the bias towards only beneficial techniques being adopted.

Causes of Change

In the case considered above, where the only mechanism of change was search, all change in the environment took place via individual firms replacing their old trait with a new one – sometimes copied from another firm, sometimes obtained *de novo*. In the case where the only mechanism was 'capital change', all change was of firm *size*. Neither of these modes of change bears much superficial resemblance to reproduction and inheritance; both of these modes nonetheless serve the same function that reproduction and inheritance do in biological models of ENS. For ENS to occur, relative success must influence the subsequent prevalence of the traits that vary with regard to success. Reproduction, growth, and 'propensity of adoption',[25] all are ways that the prevalence of traits can change over time, and therefore are ways that ENS can operate.

This suggests that constructing Darwinian models doesn't require careful mimicry of the biological apparatus, finding analogues for mutation and selection, phenotype and genotype, reproduction and heredity.[26] What is required is a population, some traits, mechanisms of change, and the right parameters. The 'right parameters' are settings that allow for success[27] to cause differences in the spread of the traits. For example, in Nelson and Winter's model, if firms change their production technique in every time step, the 'capital change' mechanism can't do anything to link the profitability of a technique to its prevalence. If profitable techniques don't tend to 'stick' in the population one way or another, there can't be selection for profitability.[28] Similarly, for natural selection to operate in a consistent direction over a period of time, the influence of the environment must be sufficiently consistent that a trait remains successful over that time.

---

[25]That is, the tendency of a trait to be lost by those who have it and gained by those who don't.

[26]Nelson argues for the same conclusion on different grounds in his 2007 paper "Universal Darwinism and Evolutionary Social Science" [20]

[27]Of course, what constitutes 'success' will depend on the model – and if nothing at all can cause a type to increase its prevalence over other types, then there is no 'success' in the model in an evolutionary sense. What constitutes success is set by the mechanisms, so the identification of 'success' with profitability in Nelson and Winter's models is licensed precisely because of its role in determining firm growth and 'search' patterns.

[28]Heritability is a way of ensuring the appropriate degree of 'stickiness' when reproduction is the mode of change, and is only specifically required for such models.

Conclusion

Nelson and Winter's evolutionary models are both reminiscent of and very different from evolutionary models in biology. The analysis I have here provided conceives their models in terms of the interaction of distinct mechanisms of change that jointly produce evolution by natural selection. Seeing them in this way allows us to recognise the distinct but complementary way in which the mechanisms of capital change and search contribute to the overall predictable change in the distribution of production routines in the population. The explanation of this change formally parallels the explanatory structure of Darwinian explanations of evolution by natural selection, which shows that the analogy Nelson and Winter tried explicitly to use as a guiding principle of their model building does carry through into the substance of their models. However, consideration of Nelson and Winter's work does also demonstrate some key differences that illustrate ways in which the biology-centred conception of evolution by natural selection needs to be revised. It suggests that natural selection need not involve reproduction in any sense – a highly controversial claim within the philosophy of biology. It highlights that selection is not a specific evolutionary mechanism, contrary to Sober [1] and Stephens [18], but rather that the environment determines when a mechanism selects. It illustrates how 'mutation' can be selective, and how novelty can drive evolutionary change, in ways that do not fit neatly with the conception of 'fitness' found in the philosophy of biology. By broadening our gaze from biology, we gain a valuable alternative perspective on evolution by natural selection, which can reconfigure our understanding of evolution in a fashion that may be potentially fruitful from the perspective of biology as well.

References

1. Sober, E.: The Nature of Selection. The MIT Press, Cambridge (1984)
2. Sober, E.: Models of cultural evolution. In: Griffiths, P. (ed.) Trees of Life: Essays in the Philosophy of Biology, Australasian Studies in the History and Philosophy of Science. Kluwer Academic Publishers, Dordrecht/Boston (1991)
3. Dawkins, R.: The Selfish Gene. Oxford University Press, New York (1976)
4. Aldrich, H.E., Hodgson, G.M., Hull, D.L., Knudsen, T., Mokyr, J., Vanberg, V.J.: In defence of generalized Darwinism. J. Evol. Econ. **18**, 577–596 (2008)
5. Foster, J.: The analytical foundations of evolutionary economics: from biological analogy to economic self-organisation. Struct. Change Econ. Dyn. **8**, 427–451 (1997)
6. Witt, U.: Bioeconomics as economics from a Darwinian perspective. J. Bioecon. **1**(1), 19–34 (1999)
7. Hull, D., Langman, L.R., Glenn, S.: A general account of selection: biology, immunology, and behaviour. Behav. Brain Sci. **2**, 511–528 (2001)
8. Lewontin, R.C., Fraccia, J.: Does culture evolve? Hist. Theory **8**, 52–78 (1999)
9. Sanderson, S.K.: Evolutionism and Its Critics. Paradigm Publishers, Boulder (2007)
10. Kimura, M.: Stochastic processes and distribution of gene frequencies under natural selection (1955). In: Population Genetics, Molecular Evolution, and the Neutral Theory: Selected Papers. The University of Chicago Press, Chicago (1994)

11. Fisher, R.A.: The Genetical Theory of Natural Selection, 2nd edn. Dover Publications, New York (1958)
12. Darwin, C.: On the Origin of Species. John Murray, London (1859)
13. Matthen, M., Ariew, A.: Two ways of thinking about fitness and natural selection. J. Philos. **49**(2), 55–84 (2002)
14. Godfrey-Smith, P.: Conditions for evolution by natural selection. J. Philos. **54**(10), 489–516 (2007)
15. Lewontin, R.C.: The units of selection. Annu. Rev. Ecol. Syst. **1**, 1–18 (1970)
16. Metcalfe, J.S.: Evolutionary approaches to population thinking and the problem of growth and development. In: Dopfer, K. (ed.) Evolutionary Economics: Program and Scope. Kluwer Academic Publishers, Dordrecht/Boston (2001)
17. Saviotti, P.P.: The role of variety in economic and technological development. In: Saviotti, P. P., Metcalfe, J.S. (eds.) Evolutionary Theories of Economic and Technological Change: Presents Status and Future Prospects. Harwood Academic Publishers, Chur (1991)
18. Stephens, C.: Selection, drift, and the "forces" of evolution. Philos. Sci. **71**, 550–570 (2004)
19. Walsh, D.: Book keeping or metaphysics? The units of selection Debate. Synthese **138**, 337–361 (2004)
20. Nelson, R.: Universal Darwinism and evolutionary social science. Biol. Philos. **22**(1), 73–94 (2007)
21. Nelson, R., Winter, S.: An Evolutionary Theory of Economic Change. The Belknap Press of Harvard University Press, Cambridge (1982)

The Ethical Treatment of Animals: The Moral Significance of Darwin's Theory

Rob Lawlor

Introduction

Evolutionists are frequently required to defend themselves against the claim that Darwin's theory of evolution is immoral and that, if Darwin's theory was true, this could justify the rejection of the welfare state, or could justify an attempt to create a superior race by means of eugenics and genocide.

The standard response to this is to stress that Darwin's theory of evolution is not a moral theory, and that a moral theory cannot be derived straightforwardly from a scientific theory.

Although the concerns that Darwin's theory can justify inequality or genocide are not justified, it is true that there have been attempts to build ethical theories, or ethical judgements, on the theory of evolution.

Simon Blackburn, in the *Oxford Dictionary of Philosophy*, characterises evolutionary ethics as "the attempt to base ethical reasoning on the presumed facts about evolution... The premise is that later elements in an evolutionary path are better than earlier ones; the application of this principle then requires seeing western society, *laissez-faire* capitalism, or some other object of approval, as more evolved than more 'primitive' social forms." And he states, "Neither the principle nor the applications command much respect."[1, p. 128]

Those who argue against Darwin's theory by appealing to what they consider to be its ethical implications make the mistake of thinking that evolutionary ethics provides the correct account of the relation between evolution and ethics. They then conclude, on the basis of this, that this presents a problem for the theory of evolution. Instead, they should reject evolutionary ethics and recognise that we *should not be trying* to derive an ethical theory from the theory of evolution.

R. Lawlor (✉)
Inter-Disciplinary Ethics Applied, University of Leeds, LS2 9JT Leeds, UK
e-mail: r.s.lawlor@leeds.ac.uk

M. Brinkworth and F. Weinert (eds.), *Evolution 2.0*, The Frontiers Collection, DOI 10.1007/978-3-642-20496-8_11, © Springer-Verlag Berlin Heidelberg 2012

Nevertheless, Darwin's theory is not morally insignificant either. In this paper, I consider the ethical treatment of animals, and demonstrate how Darwin's theory is relevant to the debate, particularly in posing problems for those who wish to defend speciesism.

Speciesism

Peter Singer and Michael Tooley, among others, have argued that treating humans differently from other animals, just because they are human, is speciesist, and is comparable to racism. Both involve giving preferential treatment to individuals of one group over those of another on the basis of characteristics (skin colour or species membership) that are morally irrelevant: "Difference in species is not per se a morally relevant difference." [2, p. 51].

Tooley and Singer both argue that, instead of focusing on the species of an individual, we should consider their individual capacities. Clarifying his position, Singer insists that a rejection of speciesism does not commit one to the view that all life is equally valuable. He writes:

> When we come to consider the value of life, we cannot say... confidently that a life is a life, and equally valuable, whether it is a human life or an animal life. It would not be speciesist to hold that the life of a self-aware being, capable of abstract thought, of planning for the future, of complex acts of communication, and so on, is more valuable than the life of a being without these capacities [3, p. 61].

He does argue, however, that it would be speciesist to treat one life as being more valuable than another if this was based *solely* on a preference for members of our own species – for example, in a case in which we are considering the life of a human and the life of a non-human animal, *where the capacities of the two are equal.*

For example, Singer asks us to consider the case of medical research. Singer concedes that it is possible to appeal to the different capacities of animals and humans in order to argue that it is better to experiment on animals than it is to experiment on normal adult humans, but he goes on to claim that the argument also gives us reason to experiment on "severely intellectually disabled humans for experiments, rather than adults." [3, p. 60]

For those not familiar with Singer's view, it is important (both for clarity and for fairness) to emphasise that Singer has always claimed that this is not meant to be an argument in favour of experimenting on the severely intellectually disabled, but an argument *against* experimenting on animals.

Singer then asks:

> If we make a distinction between animals and *these* humans, how can we do it, other than on the basis of a morally indefensible preference for members of our own species? [3, p. 60]

In response, some people have embraced the label, stating boldly that they are speciesist (though typically denying the comparison with racism). Carl Cohen, for example, states "I am a speciesist. Speciesism is not merely plausible, it is essential for right conduct." [4] And LaFollette and Shanks observe that "Most researchers now embrace Cohen's response as part of their defence of animal experimentation." [5] Similarly, David Oderberg writes: "the view I am defending seeks to preserve an essential difference between [animals and humans]" [6, p. 140].

The aim of this paper is to draw on the details and implications of Darwin's theory of evolution in order present a challenge which will need to be met by those who wish to defend speciesism.

In the remainder of the paper, I will present two different ways in which we can characterise speciesism – the moral status interpretation and the relational interpretation – and will argue that, on either characterisation, there are difficulties for those defending speciesism. I don't necessarily see these as providing a conclusive rejection of speciesism. However, I do consider them to be important challenges that need to be met by Cohen and others who want to embrace speciesism.

First though, in order to clarify my position, I want to distinguish it from a different argument which appeals to evolution in order to pose problems for speciesism.

A Common Argument: An Appeal to Common Ancestry

One common way of arguing against speciesism is to appeal to the common ancestry of humans and other animals, and to argue, on these grounds, that we cannot justify our treatment of nonhuman animals.

Peter Singer, for example, states that we ought to:

> Recognise that the way in which we exploit nonhuman animals is a legacy of a pre-Darwinian past that exaggerated the gulf between humans and other animals, and therefore work towards a higher moral status for nonhuman animals, and a less anthropocentric view of our dominance over nature [7, pp. 61–62].

But this argument from Singer seems to conflict with his main arguments against speciesism. If the claim is that we should treat animals better in virtue of our common ancestry, then this approach will share the same features that are supposed to make speciesism morally problematic. Opponents of speciesism, like Singer, ask: why should I treat one animal better than another just because one happens to be the same species as me? Similarly, we can ask Singer, why should we treat animals better than we did in the past just because we now recognise the common ancestry that humans share with the other animals? Why is this relevant? If preferential treatment for members of our own species is impermissible then surely the preferential treatment for those with a common ancestry (which we might call ancestorism) should be impermissible too.

I doubt, therefore, that an appeal to our common ancestry will help us to reject speciesism. Indeed, this ancestorism would seem to support speciesism. (I defend this claim in more detail later in the paper.)

The Moral Status Interpretation of Speciesism

On the first interpretation, speciesism states that different species have different moral status. On this account, in its simplest form, any human will have the moral status appropriate for humans, while any dog will have the moral status appropriate for dogs, regardless of the individual capacities of the individual human or the individual dog.

On this account, the moral status an individual has is based on the species it belongs to, rather than on the particular characteristics of the individual. The speciesist, in contrast to Singer, claims that the individual capacities of each individual are not important (or, at least, that the individual capacities are not the only important consideration). Rather, the moral status each individual has is dictated (primarily) by the species it belongs to.

At this point, it may simply look like a clash of intuition against intuition. Singer claims that any appeal to the species of the individual is irrelevant, and therefore comparable to racism, while the speciesist insists that species membership is a morally significant consideration, and that a human has a particular moral status simply in virtue of being human.

Often the latter view will be informed by religious beliefs, but it needn't be. It may, instead, be informed by the intuition that experimenting on the severely intellectually disabled infant is worse than experimenting on a non-human animal. Or it may be informed by a particular argument – for example, Oderberg's argument that the "*kind* of thing an entity is determines its potentialities" [8, p. 179] and that potentiality is important even in cases where the potential can't be achieved ([8], p. 181 and Sect. 4.4).

However, if speciesism is expressed in terms of moral status, the speciesist will need to give an account of speciesism that is compatible with a proper understanding of evolution and of what it means to say that two animals belong to the same species.

Moral Status and Transitivity

The concept of transitivity is easier to understand if it is explained using examples. In his *Dictionary of Philosophy*, Simon Blackburn defines a transitive relation as follows: "A relation is transitive if whenever Rxy and Ryz then Rxz." [1, p. 380] That is, the relation, R, is transitive if it is the case that: whenever there is a certain relation, R, between x and y, and there is also the same relation, R, between y and z as well, then it *must* be the case that there will be the same relation, R, between x and z. To give an example, the relation of being "taller than" is transitive: if Jane is taller than Jack and Jack is taller than Fiona, then Jane *must* be taller than Fiona. The relation of being "fond of", on the other hand, is non-transitive: if Jane is fond of Jack and Jack is fond of Fiona, it needn't be true that Jane is fond of Fiona.

So how does this relate to our treatment of animals? The problem for those who want to embrace speciesism is that the relation of having the same moral status should be transitive. That is, if A has the same moral status as B, and B has the same moral status as C, then A ought to have the same moral status as C. The reason this is a problem is because, once we understand what it means to say that A and B belong to the same species, we will see that moral status would not be transitive if it was determined according to species membership.

Both Dawkins [9] and Darwin [10] claim that the separation of animals into distinct species is only possible because of the *merely contingent* fact that the intermediates have become extinct. If we considered all the animals that had ever existed, we would not be able to identify distinct species. Rather, we would simply have a smooth continuum of animals, without any clear point at which we could separate one group from another, and identify them as separate species. "When we are talking about all the animals that have ever lived, not just those that are living now, evolution tells us there are lines of continuity linking literally every species to every other." [9, p. 317].

It is important, however, to stress that it is *not* the difficulty of finding a cut-off between species that is at the heart of *my* objection. The difficulty in finding a cut-off could just be a problem of vagueness. The problem I want to draw attention to is the *non-transitivity* of species membership.

Dawkins states that "Non-interbreeding is the recognised criterion for whether two populations deserve distinct species names." [9, p. 309].

The smooth continuum between all the animals that have ever lived, combined with the definition of what it is for two populations to deserve distinct species names, poses a problem for the speciesist who understands speciesism in terms of moral status.

Dawkins considers the smooth continuum between all the animals to present a problem for "the essentialist mind". [9, p. 318] When Dawkins refers to the essentialist mind here, he is referring to those who think that there is such a thing as the essence of a human, or of any other species: the idea that "Hanging somewhere in ideal space is an essential perfect rabbit, which bears the same relation to a real rabbit as a mathematician's perfect circle bears to a circle drawn in the dust." [9, p. 318].

As an example of essentialism in relation to speciesism and the special status of humans, consider the following passage from Oderberg:

> ...when Aristotle said 'man is a rational animal', he was not making a statement about only those mature, normally functioning members of humankind, when they are awake, and not drugged, and not insane, and thinking clearly, and forming plans, and making choices about what sort of life to live. He was defining the *essence* of humankind, in other words, he was telling us what human nature is, and hence what *every* human being is, simply by being a member of humankind... Why, then, should being immature or damaged (for instance) detract from the moral status of certain human beings if they are, by their very nature, every bit as human as their mature, normal fellows? [8, pp. 82–83]

Dawkins imagines having a time machine, going back in time 1,000 years at a time, picking up a young and fertile passenger at each stop. Each individual,

Dawkins asserts, would be able to interbreed with a predecessor of the opposite sex from the next stop, 1,000 years earlier. Dawkins writes:

> The daisy chain [of ancestors] would continue on back to when our ancestors were swimming in the sea. It could go back without a break, to the fishes, and it would still be true that each and every passenger transported 1,000 years before its own time would be able to interbreed with its predecessors. Yet, at some point, which might be a million years back but might be longer or shorter, there would come a time when we moderns could not interbreed with an ancestor, even though our latest one-stop passenger could. At this point we could say that we had travelled back to a different species.
>
> The barrier would not come suddenly. There would never be a generation in which it made sense to say of an individual that he is *Homo sapiens* but his parents are *Homo ergaster* [9, p. 319].

Here, Dawkins stresses the vagueness and the smooth continuum between species, arguing that there isn't a clear point at which we can say this is a human but this predecessor is not. But, from the point of view of demonstrating the difficulty of determining moral status by species membership, there is a more important lesson to take from this journey back in time.

Consider Dawkins' statement that: "At this point we could say that we had travelled back to a different species." Now consider one of our time travelling passengers – one that we picked up before we got back to a different species (at our 500th stop, for example). I will call him "Grunt". At the point at which we could say that *I* had travelled back to a different species, Grunt wouldn't have. Grunt would still be able to interbreed with the predecessors while I couldn't. Call the passenger that we pick up at this stop "Ugh".

Thus, it seems that Grunt and I are the same species, yet Grunt is also the same species as Ugh, even though Ugh is a different species from me.

This journey back in time demonstrates that (on the interbreeding criterion) it is perfectly possible for A to be the same species as B, and for B to be the same species as C, but for A to be a different species from C. Species membership, on this account, is non-transitive.

The problem is that moral status should be transitive. That is, if I have the same moral status as Grunt, and Grunt has the same moral status as Ugh, then I must have the same moral status as Ugh. But species membership, unlike moral status, is not transitive. Therefore, moral status cannot be determined by species membership.

I should stress, at this point, that this is not meant to be an objection based on actual *practical* difficulties in identifying distinct species. As Dawkins says, we are actually very lucky in that respect.

> Creationists love 'gaps' in the fossil record. Little do they know, biologists have good reason to love them too. Without gaps in the fossil record, our whole system for naming species would break down [9, p. 319].

As this quote suggests, the problem is not a practical one. We can (usually) identify different species. Rather, the objection is simply that, if we consider the implications, it doesn't look like moral status should be determined in this way.

The force of this objection will become clearer if we consider ring species.

Ring Species and Taxonomy

Above, I argued that the criterion for "being the same species as" is non-transitive. Taxonomists, however, want an account of species membership that is transitive and allows us to sort animals into categories. That is, the point of talking about species is not to make relational statements, such as A and B are able to interbreed with each other. Rather, the aim is to *sort animals into groups* – according to species membership.

Ultimately, the criterion that taxonomists use to sort animals into groups is actually very poorly suited to this task. In most cases, this doesn't cause any problems in practice. In some cases, however, there is a complication. Dawkins describes a type of Ensatina in California, which is an intermediate between "two clearly distinct species of *Ensatina* which do not interbreed": [9, p. 309].

It is not a hybrid. That is the wrong way to look at it. To discover the right way, make two expeditions south, sampling the salamander populations as they fork to west and east on either side of the Central Valley. On the east side, they become progressively more blotched until they reach the extreme of the *klauberi* in the far south. On the west side, the salamanders become progressively more like the plain *eschscholtzii* that we met in the zone of overlap at Camp Wolahi.

This is why it is hard to treat *Ensatina eschscholtzii* and *Ensatina klauberi* with confidence as separate species. They constitute a 'ring species'... Zoologists normally follow Stebbin's lead and place them all in the same species [9, pp. 309–310].

Ultimately, this is an *ad hoc* solution to the problem that that the criteria used to identify species membership are not actually well-suited to sorting things into groups. For the zoologist, this *ad hoc* solution serves the purpose adequately. For the ethicist seeking a morally significant difference between humans and non-humans the problem is much greater.

The Implications of These Thoughts About Ring Species

At first glance, it might look like the taxonomists' understanding of species might be helpful to the moral status speciesist. Like the moral status speciesist, the taxonomist wants an account of species membership that is transitive and allows us to sort animals into distinct groups. Therefore, if the taxonomist is able to solve this problem, the speciesist can help himself to the same solution.

The problem with this approach, however, is that it has implausible implications. In particular, on this account, the moral status of a particular type of animal will depend on contingent and seemingly irrelevant facts about which other animals have become extinct. Consider the case of the Californian salamanders. If the intermediate salamanders had become extinct, taxonomists would have regarded the *Ensatina eschscholtzii* and *Ensatina klauberi* as separate species, but because the intermediates do still exist, they are treated as a ring species.

To see why this approach is problematic for the ethicist who wants to link moral status to species membership, consider the following quote from Dawkins:

> Ring species like the salamanders and the gulls are only showing us in the spatial dimension something that must always happen in the time dimension. Suppose we humans, and the chimpanzees, were a ring species. It could have happened: a ring perhaps moving up one side of the rift valley, and down the other side, with two completely different species co-existing at the southern end of the ring, but an unbroken continuum of interbreeding all the way up and back round the other side [9, pp. 311–312].

According to the (moral status) speciesist, chimpanzees don't have the same moral status as humans, and this is because they are a different species. However, if the speciesist adopts the taxonomist's strategy of appealing to ring species in order to avoid the problem of non-transitivity, speciesism begins to look absurd. If we ask the speciesist what he would say about the moral status of chimpanzees if the scenario described above was a reality, the speciesist would have to concede that, because (in the scenario we are imagining) humans and chimpanzees are the same species (a ring species like the Californian salamanders), they must have the same moral status.

My claim is not that it is absurd to say that chimpanzees should have the same moral status as humans. What is absurd is not the conclusion but the method of reaching the conclusion, and the fact that the moral status of chimpanzees would depend on whether or not there is a living continuum between chimpanzees and humans in the way that Dawkins describes.

Imagine that there is a living continuum between chimpanzees and humans, such that the speciesist who embraces the concept of ring species is required to say that they both have the same moral status. Now imagine that every single one of the intermediate animals is killed in a massacre, such that there is no longer a continuum between humans and chimpanzees. Now, it seems that, as a result of this massacre, humans and chimpanzees could now be recognized as separate species and, as a result, the chimpanzees would lose the special moral status they once had in virtue of being placed in the same species as the humans.

The thought that the moral status of chimpanzees could change is this way is clearly absurd. Of course, the speciesist can avoid this particular absurdity by refusing to accept ring species, but then of course they have the problem we started with.

Definitions of "Species"

I quoted Dawkins earlier, saying that "Non-interbreeding is the recognised criterion for whether two populations deserve distinct species names." [9, p. 309] However, this is not entirely uncontroversial. Biologists argue about the best criteria for distinguishing between species. But the problem I have identified is not restricted to one particular definition of "species".

Most dictionary definitions[1] state three criteria: physical resemblance, close relation and the ability to interbreed. And, indeed, Dawkins too does seem to share this fuller understanding of "species". See, for example, his discussion of salamanders and gulls in "The Salamander's Tale" [9, pp. 308–320].

Regarding the criterion of being closely related, I take it that the ability to interbreed is considered important primarily because if two populations are capable of interbreeding this is evidene that they are closely related.[2]

Therefore, even if we continue to take non-interbreeding to be the *criterion* for deciding whether two populations deserve distinct species names, this needn't be because interbreeding is – in itself – important. Rather, two populations deserve to be grouped together as a single species if they are sufficiently closely related. Interbreeding is just a way of *identifying* that they are closely related.

As far as I can see though, this clarification of the understanding of what it is for two populations to be grouped together as a single species does not make any significant difference to the plausibility of speciesism, on the moral status interpretation of speciesism. If speciesism is based on an understanding of an account of species that relies on physical resemblance or being closely related, speciesism will have the same problems we saw before. Species membership will still be non-transitive. A can resemble B and B can resemble C, but A needn't (sufficiently) resemble C.[3] Likewise (according to the relevant standard), A can be closely related to B and B closely related to C, but A needn't be (sufficiently) closely related to C.

Regardless of whether taxonomists use the resemblance criterion, the close relation criterion or the interbreeding criterion, or some combination of these, "being the same species as" will still be non-transitive – unless we appeal to the concept of ring species – and the problem will remain.

A Position Very Similar to Moral Status Speciesism

Despite this, it might be possible to defend a position very similar to moral status speciesism, but appealing to something other than species. If we could sort animals and humans into groups according to something other than species membership, then we might be able to construct an alternative to speciesism.

Of course, this would involve a rejection of speciesism, but this may nevertheless be a result that speciesists would be happy with. If such a position could be defended, this would be much more likely to be seen as a victory for the speciesist than for their opponents like Singer. Thus, if you are persuaded by Oderberg's arguments that moral status should be conferred to kinds rather than to individuals,

---

[1]Based on a brief search of half a dozen dictionaries.

[2]And, of course, the interbreeding criterion can't be the criterion for life forms that reproduce asexually.

[3]See Dawkins' examples of the salamanders and the gulls [9, pp. 308–311].

([8], Sect. 4.4) my arguments suggest that this shouldn't be done according to species membership, but *not* that it can't be done at all. My claim is simply that membership to these groups or kinds will need to be transitive, and species membership isn't. As such, those who want to group animals into different kinds, in order to confer moral status according to these kinds, will need to sort them according to something other than species membership. What these groups or kinds could be, however, I do not know.

The Relational Interpretation of Speciesism

Now though, I will consider the possibility that speciesism could be defended by rejecting the moral status interpretation of speciesism and offering an alternative understanding of speciesism, which doesn't rely on the claim that humans have a special moral status not shared by other animals, and doesn't therefore rely on the claim that species membership must be transitive.

The way in which relational speciesism can avoid problems associated with the non-transitivity of species membership is best explained by analogy.

Consider the case of half-brothers, and let's stipulate, as a premise for our argument, that one has special obligations to one's family, and that this includes half-brothers and half-sisters aswell as full siblings.

The relation of being a half-brother is non-transitive. The following example, therefore, is perfectly possible: John is Jack's half-brother, Jack is Stephen's half-brother, but John isn't Stephen's half-brother. As such, this allows for the non-transitivity of obligations to others on the basis of family: John has special obligations to Jack, and Jack has special obligations to Stephen, but John *doesn't* have special obligations to Stephen.

It should be noted though that this has nothing to do with moral status. John has special obligations to Jack, but not to Stephen, but the claim is *not* that Stephen has lower moral status than Jack. The obligations are based on the relations between the individuals, and this is perfectly consistent with the claim that John, Jack and Stephen all have the same moral status.

Thus, if we offer an analogous account of speciesism, the claim is that we have special obligations to those who belong to our species. But this isn't put in terms of moral status, but rather in terms of special obligations. On this account of speciesism, the claim is not that all humans have a higher moral status than other species regardless of their individual capacities, but just that we have special obligations to our own species.

I should stress, however, that it is possible to be a moral status speciesist *and* a relational speciesist, or one or the other, or neither. For example, if we imagine that Martians exist, and that they are rational and intelligent and so on, someone could think that Martians have the same moral status as humans (in virtue of the fact that – as a kind – they are intelligent, rational, self-conscious etc.), and that animals have a lower moral status, such that certain harms, which would be permissible if inflicted on a dog, would be impermissible if inflicted on a Martian (just as they would be impermissible if inflicted on a human). At the same time, however, they could also believe that it is permissible (perhaps even obligatory) for humans to give preferential treatment (in job interviews for example) to other humans (over Martians) and likewise for Martians to give preferential treatment to other Martians (over humans), basing this on the idea of special obligations, and not on differences in moral status.

Ancestorism and Speciesism

It is difficult to see how one would defend speciesism without defending ancestorism, or *vice versa*. Rather, both seem to be based on the same basic principle: that it is permissible (and/or obligatory) to give preferential treatment to those we are most closely related to.

If this is the case, though, rather than stating that people have a special obligation to anyone of the same species, and insisting that the principle of giving preferential treatment to those we are closely related to applies only at the level of species membership, it seems more plausible to suggest that we have a *range* of special obligations (that vary in degree) such that the more closely related we are to another creature, the greater our obligations to that creature.

It should be noted, however, that this is not a rejection of speciesism, opting for ancestorism instead. Rather, this form of gradable ancestorism will entail speciesism: I am more closely related to members of my own species than I am to members of any other species, and therefore I have stronger special obligations to my own species than I do to members of any other species.

The First Challenge for the Relational Interpretation of Speciesism

The original objection to speciesism was that, like racism, it appealed to something that was morally irrelevant. As such, a key objection was that speciesism was comparable to racism. On the relational account, this seems especially problematic. First, on this account the speciesist is *explicit* in saying that I should give preferential treatment to some beings, but not others, just in virtue of the fact that I am more closely related to them, or more closely resemble them. As such, the analogy with racism seems much clearer.

Second, and more worryingly, speciesism may not just be analogous to racism. Rather, a commitment to speciesism may also *entail* a commitment to racism. If we accept the suggestion above that ancestorism is the foundation of speciesism (and that speciesism is just one form of ancestorism), we have to recognise that ancestorism doesn't only entail speciesism, it also entails racism.

At this point it is necessary to qualify this claim in two ways.

First, it doesn't entail what we might call moral status racism – the view that some races have higher moral status than others. This, for example, is the racism of white supremacists and Nazis. Clearly, this would be an unpalatable conclusion. As such, it is a good thing for relational speciesism that it does not entail *this* sort of racism. Nevertheless, relational speciesism *does* entail relational racism: the view that, although there is no reason to think that different races have a different moral status, it would nevertheless be permissible (perhaps even required) for members of race A to give preferential treatment to other members of race A, and for members of race B to give preferential treatment to other members of race B etc. This doesn't seem to be as unpalatable as moral status racism, but nevertheless relational racism is still a form of racism and a view that many will be eager to avoid.

However, maybe some people – emphasising that it is not moral status racism – will accept that the ancestorist principle entails a form of racism, perhaps arguing that it is a relatively benign form of racism.

A note about this discussion of racism: At this point I should acknowledge that it may be controversial to claim that ancestorism entails racism because even the concept of race is controversial, when applied to humans. Dawkins cites R.C. Lewontin as stating that the differences between the different races are so small, compared with the differences between individuals within any particular race, that race has no meaning, as applied to humans, and that racial classification should be seen "to be of virtually no genetic or taxonomic significance" [9, pp. 417–418].

Thus, we might reasonably conclude that, if the concept of race is a confused concept and has no biological foundation, then racism too will have no biological foundation, and ancestorism will not entail racism. However, I think there are two reasons to resist this conclusion.

First, it is not clear that Lewontin is right. Following A.W.F. Edwards, Dawkins argues that we shouldn't be looking at the *levels* of variation between groups, but at the extent to which certain characteristics *correlate* with other characteristics within a racial group.

As such, it may be true that there is very little genetic difference between, for example, a Chinese man and a Nigerian, and the differences that there are may be (in most cases) differences only in appearance, but it does not follow that these differences are not racial differences.

Second, even if Lewontin is right, and Dawkins and Edwards wrong, it does not follow that we don't need to worry that a commitment to speciesism entails a commitment to racism. It is only contingently true that the differences between races are small, compared to the differences within races. Although it would not have the same practical implications, we would still have good reason to reject speciesism on the grounds that the speciesist *would* be committed to being a relational racist in a *hypothetical* situation in which the racial differences were genetically significant, such that it wasn't a mistake to recognise different races.

When a white employer gives a job to a white candidate, rather than a black candidate who is better qualified for the job, or when a government implements a form of apartheid that discriminates against (a group it perceives to be) a racial minority, the most important objection is *not* that the employer or the government has failed to recognise that the human population bottlenecked sometime in the past, maybe 70,000 years ago, with the consequence that there is "an unusually high level of genetic uniformity in the human species, despite superficial appearances." [9, p. 416] The more important objection is simply that *they are racist*: that they give preferential treatment to some, over others, on the basis of characteristics that are morally irrelevant.

As such, a rejection of racism should not depend on who is right about the *genetic* significance of racial classification: Lewontin or Edwards. See [9, pp. 415–425] and [11].

Ancestorism, Speciesism and Racism

Regarding the principle that we have special obligations to those we are most closely related to, most of us have what appear to be conflicting intuitions.

In relation to our immediate family, most people accept that we have special obligations to our parents, children and siblings, and this doesn't look like an embarrassing implication of the principle.

However, when we move to more distant relations to consider different races, many want to resist the conclusion that we have special obligations to members of our own race and that we should give preferential treatment to members of our own race. For many, this does seem to be an embarrassing conclusion, which we should want to avoid.

When we move a bit further out again, however, many people do want to embrace speciesism, and therefore (if we consider the principle solely in relation to speciesism, ignoring the implications regarding race) the principle that we give preferential treatment to our closest relations looks plausible here too.

If we take a further step back, though, to consider animal classes, rather than species, we seem to have different intuitions again. If I suggested that we had different obligations to mammals than to reptiles, because of the closer relationship between humans and other mammals, compared with humans and reptiles for example, I imagine that many would consider the suggestion to be eccentric. Why should I give preferential treatment to a mouse, over a crocodile, just in virtue of the fact that the mouse, like me, is a mammal?

The problem now is that it seems that we are want to be selective in when we apply the principle that we should give preferential treatment to those to whom we are more closely related. On this account, the principle seems to be relevant at one level (family), then irrelevant (race), then relevant again (species), and then irrelevant again (animals classes). This appears to be in need of explanation.

Although it may have unpalatable implications, it would seem to be more consistent to appeal to the closeness of relations at the level of race as well as at the level of family and species.

If you want to argue *against* speciesism, on the other hand, the same problems don't arise. There seem to be two options: first, you could argue that the principle of having special obligations to those more closely related to you applies only to those very closely related, giving one obligations to one's own family. However, these obligations would not extend any further than that and therefore would not justify giving preferential treatment to your own race over others, or your own species over others, or your own class of animal over others.

Alternatively, you could reject the principle completely and argue that the obligations we have to family are based on the social relationship rather than the genetic one (or, more radically, you could argue that we don't have special obligations to our families at all).

Either way, the position seems to be more coherent than the speciesists'.

If you want to defend speciesism, you could try to explain why the principle seems to turn on and off, being on at the level of close relations of family, turning off at the more distant relations of belonging to the same race, but turning on again when the relations become even more distant, considering species, but turning off again at the level of considering classes. How you would do this, however, is not clear.

Alternatively, if you are not able to provide an explanation, the other option is to accept that the principle does apply to the different levels, and to accept that this does commit you to a form of racism.[4]

In defence of this account, however, you might argue that this form of racism is relatively benign. You might emphasise the fact that this is very different from moral status racism. You might stress that this form of racism is consistent with a commitment to racial equality and (some degree of) equality of opportunity. While it does allow race A to favour other members of race A, over race B, it also allows race B to favour other members of race B, over race A. Furthermore, there could be legitimate restrictions on this. Even those who think that we have special obligations to members of our own family can oppose some forms of nepotism – especially in cases where an individual is in a position of power. For example, I can think that the Prime Minister has a special obligation to his own children that he doesn't have to other children, but also insist that it would not be legitimate for him to fill all the major government posts with members of his immediate family. Similarly then, it may be possible to argue that a special obligation to one's own race should be limited in similar ways.

I am not convinced, however, that this form of racism is entirely benign. For the purposes of this paper, however, I do not need to resolve this issue. This is because, even if this line of argument is accepted, there is a second challenge to relational speciesism. When combined with even this weaker form of the first challenge, the

[4]If you want to resist the conclusion that we have special obligations to other mammals, over reptiles for example, this looks less problematic than in the previous case, because the idea that the principle may effectively fade away at some stage seems more plausible than the on and off nature of the previous account.

second challenge presents a significant problem for Cohen and Oderberg, and others who want to maintain an essential difference between humans and other animals, if they appeal to relational (rather than moral status) speciesism.

The Second Challenge for the Relational Interpretation of Speciesism

The second challenge is that, for those who want to argue for a *very* different treatment of animals compared to humans, it is not clear that this form of speciesism will suffice. The most basic duties, such as the duty not to kill and the duty not to inflict significant amounts of pain, do not look like the sort of thing that is likely to be grounded on a *special* duty of this kind. If we consider duties to rescue, and we consider the common example of a choice between saving a dog and saving a severely mentally disabled human (where it is not possible to save both), it is plausible to think that this argument could be used to justify saving the human.[5] In rescue cases, we typically do think that these sorts of special duties can count in favour of saving the individual to whom you have special obligations.

Typically, however, we don't think that these kinds of special obligations allow me to kill a stranger so that I can take his organs to save my brother. Rather, the obligation not to kill is based on something more substantial, which cannot be outweighed by the sort of special obligations we are considering here.

Thus, by analogy, even if we allow speciesism, in this relational sense, it is not clear what this will allow us to justify. Nevertheless, it *might* be sufficient to distinguish between animals and humans, such that we have a response to Singer's challenge regarding animal experimentation.

Singer's argument was about consistency, in relation to moral status. As such, those wanting to defend animal (but not human) experimentation could respond by accepting Singer's claim that the animal and the severely intellectually disabled human have the same moral status, and they could say that, as long as we focused on moral status alone, it would appear to be permissible to experiment on either. Nevertheless, according to the relational speciesist, what differentiates the two is that, in the case of the human, but not in the case of the animal, we have a special obligation to the human, analogous to the *special* obligation that one has to one's siblings.

As such, this account does seem to have an advantage over the moral status account: it avoids the problems that result from the non-transitivity of species

[5]The human is severely mentally disabled in this case so that the human does not have capabilities that the dog lacks, so that if we follow Singer's suggestion of judging individuals by their individual capabilities, we would not be able to distinguish between the two. As such, if we think we should save the human rather than the dog, it would seem to be *just* because he is human, and not because he is autonomous or has capacities that the dog does not.

membership, but still allows us to respond to Singer's challenge and reach the common sense conclusion that it is worse to experiment on severely intellectually disabled humans than on animals.

Nevertheless, the account still seems problematic. For someone who wants to defend a position like Oderberg's, this argument doesn't seem strong enough. We may have responded to Singer's challenge, such that we can argue that we can experiment on animals but not on humans, but it is not clear that we have preserved "an essential difference" [6, p. 140] between humans and animals in the way that Oderberg, for example, wants to. For many of Singer's opponents, it is not sufficient to avoid the practical implications of Singer's arguments. They are opposed, also, to the fact that Singer "lowers humans to the level of other animals" [6, p. 140] and they are opposed to the claim that some humans have lower moral status than others and to the claim that some humans have a moral status comparable to the moral status of non-human animals. It is important to recognise, therefore, that relational speciesism will not be sufficient to reject *these* claims about the moral status of particular humans.

The Two Problems Considered Together

It is also important to consider the first objection in relation to the second objection. The two objections pull in opposite directions. If the speciesist responds to the second objection by arguing that the special obligation is in fact more significant than I am suggesting, the concern is that this will also justify even more significant (and less benign) forms of racism too, making the first objection stronger.

If the speciesist responds to the first objection by arguing that the sort of preferential treatment on the basis of race (or similar) that could be justified by this form of ancestorism is relatively benign or insignificant, the second objection will be much stronger. As such, it is not clear that the relational speciesist will be able to respond to *both* of these criticisms.

Conclusion

My aim in this paper has not been to argue against speciesism. Although I present arguments which I believe pose real challenges to Cohen's claim that speciesism is "essential for right conduct" or Oderberg's appeal to essentialism, I find it hard to resist the common sense view that experimenting on severely mentally disabled humans would be worse than performing the same experiments on animals. I also have some sympathy for Oderberg's claim that potentialities are important and that "the kind of thing an entity is determines its potentialities".

My aim, rather, was to highlight some of the difficulties that we need to address if we are to improve how we understand the ethical issues relating to our treatment of animals, and to highlight the fact that there is something unsatisfactory about the idea that everything is much less problematic if we simply embrace speciesism. Furthermore, with particular reference to Darwin's influence, and the relation between evolution and ethics, my aim was to argue that, although we must always remember that Darwin's theory is not a moral theory (and that it should not be appealed to in order to defend a survival of the fittest approach to ethics), we should not go to the other extreme and assume that it is morally insignificant

Acknowledgements For comments and helpful advice on previous drafts, I am grateful to Paul Affleck, Mikel Burley, Daniel Elstein, Shane Glackin, Gerald Lang, Chris Megone, Georgia Testa, and an anonymous referee for this book.

References

1. Blackburn, S.: Oxford Dictionary of Philosophy. Oxford University Press, Oxford (1996)
2. Tooley, M.: Abortion and infanticide. Philos. Public Aff. **2**(1), 51 (1972)
3. Singer, P.: Practical Ethics, 2nd edn. Cambridge University Press, Cambridge (1993)
4. Cohen, C.: The case for the use of animals in biomedical research. N. Engl. J. Med. **315**(14), 867 (1986)
5. LaFollette, H., Shanks, N.: The origin of speciesism. Philosophy **71**, 41 (1996)
6. Oderberg, D.: Applied Ethics: A Non-consequentialist Approach. Blackwell Publishers, Oxford (2000)
7. Singer, P.: A Darwinian Left: Politics, Evolution, and Cooperation. Yale University Press, New haven (2000)
8. Oderberg, D.: Moral Theory: A Non-consequentialist Approach. Blackwell Publishers, Oxford (2000)
9. Dawkins, R.: The Ancestor's Tale: A pilgrimage to the Fawn of Life. Phoenix, London (2005)
10. Darwin, C.: The Origin of the Species – Wordsworth Classics of World Literature. Wordsworth Editions, Ware (1998)
11. Edwards, A.W.F.: Human genetic diversity: Lewontin's fallacy. Bioessays **25**, 798–801 (2003)

Part III
Philosophical Aspects of Darwinism in the Life Sciences

Is Human Evolution Over?

Steve Jones

There has been a long history of people having an interest in the future. Indeed much of the Old and New Testament is involved with speculating about a world yet to come, and the ancient Greeks and others had similar concerns. But the notion of the future was formalised in the English language, by the famous novel 'Utopia' written by Thomas Moore in 1516. In this book, and many others that followed the same mould, society is revolutionised. For example, chamber pots are made of gold because it is a malleable metal and useful for that; sick people are sent to prison because they cannot look after themselves whereas criminals are sent to hospital because there must be something wrong with them. These are interesting ideas: society has changed but physically, people look very much as they do today.

In modern utopias, where predictions are made about the future, the common theme is that physically people do not look like humans however, society is very much the same as it is today with warring tribes, a hierarchy, violence, love interest, crime etc. This is quite a radical shift in that our view of the future has changed from that of a social change in which the biology of the future is pretty much the same as today, to a view emphasising biological change.

The rest of the chapter will therefore involve not vertical science but genetics, evolution and biology. This change in view happened about 100 years ago with HG Wells *The Time Machine* (1895). This is considered to be the first book with a modern science fiction plot. A time traveller arrives in a town in the future and meets the Eloi who are charming, bourgeois intellectuals. As the plot develops it turns out that these people have a terrible secret. Nearby live the Morlocks. They are violent, drunken hooligans who live underground and regularly come out at night and kill and eat the Eloi. What has happened is that the human race has split or evolved into two. This is very much a Darwinian view. The twist in the tale is that the Eloi are the domestic animals of the brutish Morlocks who are in fact the rulers

S. Jones (✉)
Galton Laboratory, University College London, London WC1E 6BT, UK
e-mail: j.s.jones@ucl.ac.uk

M. Brinkworth and F. Weinert (eds.), *Evolution 2.0*, The Frontiers Collection, DOI 10.1007/978-3-642-20496-8_12,

and the Eloi are allowed to live until they are needed as food. This is a pessimistic view of the future where the biology of the future is one of decline into a race of thuggish, violent hooligans and is consonant with the widespread view nowadays that the future is doomed for some reason, for example, because of the reproduction and increased prevalence of bad genes.

HG Wells was a keen Darwinian and protagonist of Francis Galton, Darwin's half cousin. Galton founded the National Laboratory of Eugenics (now called the Galton Laboratory in University College London). He was a remarkable man and did many eccentric things, such as producing a Beauty Map of the British Isles. He scored local females on a five-point scale from attractive to repulsive. Galton had a great interest in human quality and wrote the book, 'Hereditary Genius' which in some quarters is regarded as the first textbook of human genetics, which it really is not. Galton also greatly influenced racial thinking. He was a pioneer of eugenics believing that undesirable characteristics should be repressed in some way and desirable ones should be encouraged. He was convinced, with almost no real evidence, that genius and criminality were heritable and was the first person to apply statistical methods to the study of human differences.

In his racial ability diagram, (Fig. 1) Galton indicated that the ancient Greeks were more intelligent than the English, who in turn were more intelligent than Asians and Australians intelligence overlaps with dogs! This supported the field of scientific racism and the widespread view amongst scientists in those days that suggested a difference in ability between races (see [1]). Whilst this thinking did not last it is indicative of the intellectual atmosphere when 'Time Machine was written and which was behind the intellectual thinking of many modern science fiction writers.

We are now in a position to make informed guesses about the future of human evolution (unlike Galton or Moore) because so much is known about what has happened in the past regarding human evolution. Since we know about how the process of evolution works, we can speculate about where it might go in the future.

In this paper, I refer to evolution in the broad popular sense i.e. different populations becoming different from each other, not in the strict genetic sense of a change in gene frequencies.

The Darwinian argument can be summarized as follows: Evolution is descent with modification. Descent means the passage of information from one generation to the next, and modification, the fact that that passage is imperfect. It thus follows that evolution is more or less inevitable. It is an old idea and was used by linguists to understand the evolution of language. Darwin even acknowledged it was not a new idea. Darwin's words can be rephrased as: evolution is genetics plus time; where genetics refers to DNA that is being copied imperfectly because of mutation and

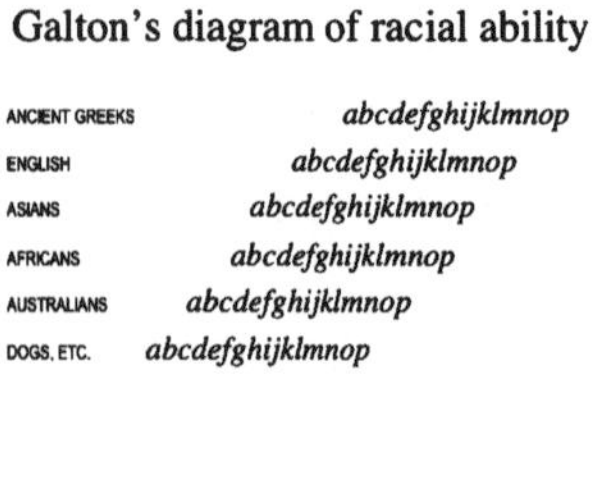

Galton's diagram of racial ability

ANCIENT GREEKS *abcdefghijklmnop*
ENGLISH *abcdefghijklmnop*
ASIANS *abcdefghijklmnop*
AFRICANS *abcdefghijklmnop*
AUSTRALIANS *abcdefghijklmnop*
DOGS, ETC. *abcdefghijklmnop*

Fig. 1 Galton's diagram of racial ability (data from [1])

time refers to more than three- and a half-thousand million years. What Darwin added to the argument is crucial and was a novel idea (it is even considered by some to be the best idea ever had). His notion was of natural selection. i.e. inherited differences leading to increased chances of reproduction. The whole Darwin machine turns on differences. Differences in genetic constitution, differences in chances of survival and differences in random accidents of time as one generation succeeds the next.

By looking what has happened in the past to three aspects of the Darwin machine: variation (mutation), natural selection and random change (random genetic drift), we can predict what might happen in the future.

Mutation

Much is known about mutagenesis and over the years there has been a real fear about radiation and chemicals ever since the 1930s when Muller showed that X-rays cause mutations. These fears are reflected in the scenarios of many science fiction books where the opening scene involves a massive radiation leak causing an increase in mutation rate resulting in the creation of horrific monsters! One such real life episode, and perhaps the most cynical scientific experiment ever, was the dropping of atomic bombs on Hiroshima and Nagaski in August 1945. This led to the end of the war and within a week a team of scientists were sent to Japan. Most were physicists curious to see what the bomb had done and were horrified by the power of it, but many were geneticists because there was a strong presumption that there would be severe genetic damage, given the massive dose of radiation to those exposed. Certainly, many people died immediately or within a month from radiation sickness because their DNA had been shattered by the huge dose of radiation. They were convinced that it was likely that the children of people exposed to the bomb would have genetic damage not seen in children of a control group not exposed to the bomb. The Atomic Bomb Control Commission (ABCC) stayed for almost 50 years although it was pretty ineffective since in 1945 little was known about human genetics. The chromosome number was not discovered until 1954, protein technology was not developed until the 1960s and DNA technology was not available in an easily usable form until the 1990s. However, towards the end of their time the ABCC did manage to detect mutations using DNA technology. They looked at millions of gene loci and tens of thousands of people and found a total of 28 mutations at the DNA level in the population as a whole [2]. There was no difference in the mutation rate of children whose parents had been exposed to the bomb compared with those who had not. Interestingly a clear pattern was seen. Of the 28 mutations found, 25 occurred in the father and only three in the mothers indicating a higher mutation rate in males than in females.

There are many other examples of increased mutation rate in males. The effect of the age of the father on mutation rates can be seen in achondroplastic dwarfism. There is a low rate of this condition in children with fathers aged less than 24, but the rate goes up over ten fold in children with fathers aged over 50 [3]. This is also true for a series of dominant (where only one copy of the gene is needed to exhibit the condition) skeletal disorders. Increased father's age has also been shown to correlate with a decrease in children's IQ. A study showed a mean IQ of 108 for children whose fathers were 18 in contrast to a mean IQ of about 100 in children with fathers aged 60.

The effect of father's age on mutation rate is due to differences in ways the sex cells are made in men and women. Every egg a woman produces is made before she is born. Eggs go through nearly all their cell division processes before birth and are then are frozen (in time). They are then released at intervals throughout a woman's reproductive life. A man on the other hand makes sperm all his adult life. Every time a sperm is made there is a further chance of error. This means there are very different numbers of cell divisions between an egg that a women passes on and the one that made her, and a sperm that a man passes on and the sperm that made him. For women there are only eight cell divisions for every egg she makes. For men the figures are very different. For a 26-year old father (the mean age of reproduction in the West) there are around 300 cell divisions between the sperm that made him and the one he passes on. In a 51-year old father, that increases to 2,000 cell divisions and in a 70 year old father that increases to 3,500. [4] Every time a cell divides there is a chance for error. This explains why there is an increase in mutations in men rather than women and why father's age has a striking difference on mutation rate. So if we want to know what will happen to mutation rates in the future we need to look at how many older fathers are there likely to be. Since the increase in mutation with age is not linear – it gets worse with age – we need to concentrate on fathers at the extreme end of the scale. It is a common belief that there are many more, older parents now than there once were, but this is not true. Across the whole world, except in Africa, society has changed its reproductive behaviour to one where people start their families later (mean age for women is 26) but they also end them earlier. Their entire reproductive lives are compressed into a short time. So actually the number of older fathers goes down as society becomes more developed.

Figure 2 shows that in Cameroon, an underdeveloped country, half of fathers are over 45. In Pakistan, a developing country where society is changing to a western lifestyle, one in five fathers are over 45 whereas in France only one in

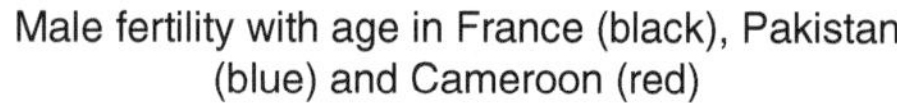

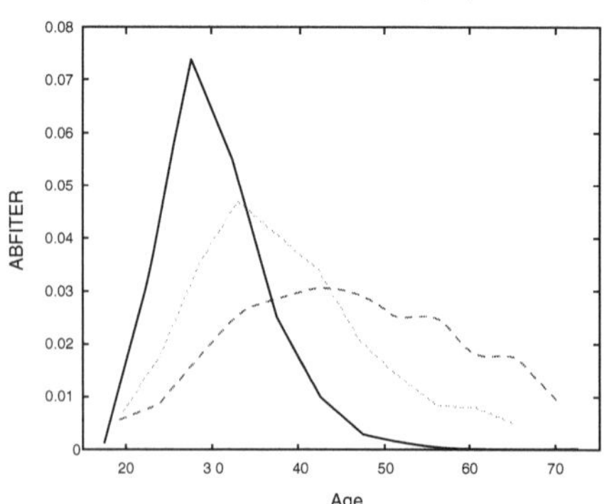

Fig. 2 Male fertility with age in France, Pakistan and Cameroon (Paget and Timaeus, 1994, cited in [5])

20 fathers are over 45 [5]. Thus, the idea that we are undergoing mutational melt down because of genetic damage due to new mutations is unlikely and may not make any difference to evolution in the future. Mutation rate is not going up it is going down.

Natural Selection

Natural selection is 'a difference in the chances of reproduction'. If an individual has a version of a gene that makes it more likely that they will survive, find a mate and reproduce, while other individuals have a version of the gene that makes it less likely, then the first version of the gene will become more common in the next generation and over time will spread allowing the population to adapt to a change in circumstances. Darwin suggested that this process may well give rise to new forms of life.

Natural selection can be thought of as a factory for making almost impossible things and is not confined to living things. For example in an actual factory making detergent, the nozzles used to make the powder frequently got blocked and were inefficient so the company tried to improve the design but with little success. What engineers turned to is a precise analogy of Darwin's natural selection. They took a nozzle and made ten copies each of which was changed slightly at random. Then they tested them against each other and against the original. If one of them did even marginally better than the others, they took that and made ten more slightly different random copies. This process was repeated again and again. Thus the nozzle began to evolve through natural selection. After 45 generations, the resulting nozzle looked quite impossible. No one had designed it but the final product worked 100 times better than its predecessor. This type of Darwinian engineering is used in the design of turbine blades and in computer science. And it works!

At first sight it might appear that there are a number of species of human beings around the world, as *Homo Sapiens* looks physically different in different regions but genetically it is striking how similar we are. The most obvious difference is skin colour. Before the mass movements of the last 300–400 years, in general, people with dark skin lived in the tropics and people with light skin lived in northern and southern parts of the world. One of the main genes involved in skin colour was found in the zebra fish (widely used in developmental biology). There is a mutant zebra fish where none of the melanin is made called the golden zebra fish. The stripes are present but they lack the black pigment. Using conventional molecular biology, the gene involved was found and fed into the huge gene database SWISSPROT, which holds information on all the genes that have ever been studied. The same gene was found in humans in two versions. Ninety-nine per cent of native Europeans have a certain protein tyrosine kinase at one particular position and 99% of all Sub-Saharans have a different protein at that position. The European version of this protein does not succeed in making melanin, the African version does. This striking difference is made by one simple change in the DNA [6].

(Incidentally Chinese/Japanese ethnic groups have the African form of the gene but also have a different mutation in the melanin producing pathway, which has similarly been selected for by natural selection.)

So white skin has evolved twice and the East Asian light-coloured skin has a different origin from European light-coloured skin. All this has happened relatively recently. The first British people to arrive over 40,000 years ago before the last ice age were probably black, so it has changed since then. Everything we know about having the gene for black melanin pigment is good. The obvious reason is that it protects from skin cancer. In pregnant women with light skin who sunbathe, folic acid and antibodies in the blood are destroyed. Dark skin is usually associated with dark eyes, which are better at seeing than blue eyes, and also better hearing because the amount of melanin in the ear is linked to the amount in skin. So if having the ancestral form of the gene is advantageous, why did light skin develop when humans left Africa?

The answer is vitamin D. Vitamin D is made through the action of UV light on 7-dehydrocholesterol in the skin. Scandinavian people who have very pale skin can make enough vitamin D to stay healthy if exposed to bright sun for only 20 min. Dark skin would not make sufficient. Deficiency of Vitamin D gives rise to the bone disease rickets. This was common in industrial cities in the nineteenth century due to a lack of sunlight. People stayed inside a lot, windows were bricked up because of the window tax, there was smog, and no oily fish in the diet, etc. However, rickets has not gone away and is still the second most common non-communicable disease of childhood worldwide. Vitamin D is also important for muscle health, blood pressure regulation and immune system as well as other bodily functions. So Vitamin D is absolutely essential and skin colour makes a tremendous difference to the amount of vitamin D that can be made.

Figure 3 shows plasma concentration of vitamin D of African Americans versus European Americans at different times of year. Most European Americans are above the desirable level of 50 while most African Americans are below, both with peaks in the summer months. Surveys of vitamin-D concentrations in Britain show that some Asian women have low levels because they are not exposed to high levels of sunshine: they wear a full covering of clothes and do not go out much. This

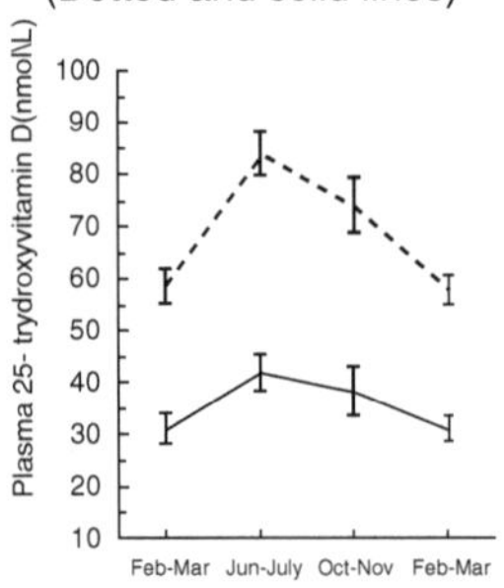

Fig. 3 Plasma vitamin D levels in European- and African- Americans (*dotted* and *solid lines*) [7]

matters less in modern world because many foodstuffs contain vitamin D. But as humans spread to areas of Europe with less sunlight. anybody without the ability to make vitamin D would have been disfavoured and the ability to make enough vitamin D because of light skin was an advantage. This led to the evolution of light skin by natural selection [8]. To take it one step further, what is the point of blond hair? Blond hair was unusual in the global context before people began to move around the world. It is known that blond hair and pale skin go together. There are about half a dozen genes involved in the blonde hair, blue eyes, pale skin package. The frequency of these genes in Scandinavia in about 1700 (before people began to move about) was about 80% of the population, in northern England was 50% and they were practically unknown in Southern European countries. So when did they turn up? To answer this we need to look to the origins of farming. Over 10,000 years ago in the fertile crescent of the middle east in what is now Iraq, new crops and grains meant the population was well fed and could reproduce and the population spread at great speed across Europe. But early grains need warm springs in order to germinate. In fact it is impossible to grow primitive strains of wheat north of a line above a latitude about parallel with Bradford. Once the spread of populations got to this latitude in mainland Europe they could not get any further. The one exception is North West Europe. Most of Europe gets warm because it has a nice, sunny springs however north west Europe gets a warm because of the gulf stream. The warmth in spring comes from the Tropics but it is accompanied by rain, thus the conditions are perfect for growing grains in February and March but there is no sun. Farming did not spread to western Europe until relatively recently. It got to the channel about 7,000 years ago, Yorkshire about 5.000 years ago. It only got to southern Scandinavia about 4,000 years ago and not to northern Scandinavia until a couple of 100 years ago. This meant there was a grain-eating population that lacked vitamin D due to low levels of sun. Under these circumstances having fair skin would be an advantage and any additional advantage such as blond hair and very pale skin would have been strongly favoured, giving early settlers a real advantage over darker skinned people. So blondness probably spread over the last few years through natural selection.

Natural selection turns on differences. If every family has the same number of children there would be no natural selection. It is only if some families have ten children and some have two and there are differences in survival that there can be natural selection. To work out the power of natural selection you do not need to know what people die of, just what the figures for life and death are. In the world today two out three people will die of a disease associated with their genetic make up. In sixteenth century two out of three babies born in Britain died before they were 21 years old. In 1800s almost one in two died and in modern Britain (2001) 99% of newborn babies will survive until they are 21 as long as they last the first 6 months. That means there are no real differences in survival rates now, so there is no natural selection. However, as Darwin recognised, natural selection has two parts. It is not just a question of survival but also the ability to reproduce. The opportunity to create differences in reproductive success is greater among men than women. Most women are limited by the facts of biology in the number of children they can have, but men are not so limited. There are many historic cases of males having many children and if some males have vast numbers of children there must be other males who have none. This effect still goes on today. Osama Bin Laden's father had 22 wives and 53 children. In communities where men have many wives then it goes without saying that some men will have no wives at all, thus giving some men many children and other men have no children at all.

So there is a massive variation in the chance for reproduction among men in that community. This is important because of the genetic information carried on the Y chromosome. In England the genetic information carried on the Y chromosome is quite varied but in other places this is not so. For example, in Donegal in Northern Ireland about 20% of men share the same Y chromosome variant. They belong to families called the O'Donnells who can trace their ancestry to the High Kings of Ireland in the fifth century. These high kings were basically warlords in that they were in charge of a violent group of men who conquered other tribes. They and their sons had many wives and mistresses. All these men can be traced back to King Niall of the Nine Hostages and it is his Y chromosome that is still around today. So the variation in reproductive success still leaves its imprint thousands of years later.

Given that there is no longer any variation in survival rate, what about variation in fertility today. It is perceived that the fertility rate has reduced however so has variation in fertility rate, i.e. most people have a similar number of children. In Western Europe the opportunity for natural selection due to reproduction variation has gone down in the last 200 years by 90% thus there can be no more Darwinian natural selection.

Evolution at Random

The third part of the Darwinian machine is Evolution at Random. Evolution can occur at random, especially in small populations. Darwin's first introduction to the ideas of change came whilst in the Galapagos Islands, where he noticed differences on different islands in the tortoises and in the mocking birds. He speculated that these birds looked a bit like the birds on the mainland of S. America but were slightly different [9]. On the Cape Verde islands the animals and birds he saw there were a bit like those on the African continent but were slightly different. He suggested that natural selection was involved because of the change in environment but also that random chance was also involved i.e. it was only by accident that some creatures got to the islands. There are always fewer species on islands compared to the mainland because by accident some creatures get there and some do not. The same is true of genes: if one looks at animals on islands they are nearly always less genetically variable than animals on the mainland and the same is true of humans.

There are plenty of opportunities for random change in small populations. In the 1870s Francis Galton first showed the power of this effect. He noticed whilst on walking holidays in Switzerland that in the isolated Italian-speaking villages, almost everyone had the same surname in one village. In another poor isolated village everyone also had the same surname but a different one from the first village. This intrigued him because he was interested in the inheritance of human quality and it appeared at first that having one surname was advantageous in one village, but having a different surname was advantageous in a different village. Then he realised it was actually an inevitable effect of genetic change in small populations. Take a village founded in say, 1300, where there were ten families with different surnames. If, in any generation a man had no sons his family name would die out and the other names would get more common. Inevitably over time and at random one name would take over and thus the information on the Y chromosome would become prevalent. This is an example of evolution at random, which is particularly important in small isolated populations.

For most of our history humans have lived in small hunter-gatherer units. Rather obviously usually the abundance of a species is related to its size. For example there are far more small mammals like the shrew than larger mammals such as the elephant. Today humans are 10,000 times more abundant than would be expected from our size, due to farming, health care, industry etc. The natural population of the world is about that of West Yorkshire. For most of human history we lived in tiny groups and in such small groups there can be rapid genetic change. If you follow the journey of humankind from its birthplace in Africa across the world to its final destination in Southern America you can see there has been a linear decrease in the amount of genetic variation. Which tells that we went through bottleneck, after bottleneck after bottleneck of small population size as we moved losing genes all the time.

This decrease in genetic variation can still be seen in some modern populations. In Finland, where traditionally people lived in isolated small communities, there are at least 30 recessive genetic diseases that are only found in Finland. An example is the recessive genetic condition VLINCL. Using the excellent family records held in Finland, a group of Finnish scientists drew up a family tree which showed that all the people with this rare disease could be traced directly (or with just one or two links missing) to a man who lived around 1650. Thus the parents of young child with the disease had no idea they were related but were actually inbred. Each carried a single copy of the rare gene and the child inherited two copies. This is a common phenomenon in small, inbred populations for example in parts of India, USA and middle East where it is common for cousins to marry.

Irrespective of medical consequences, inbreeding changes gene frequencies in populations at random. Another example is in the isolated regions of the Orkney Islands where the population is 20% more inbreed than the average Scottish population. Around the world particularly in Islamic Middle Eastern countries it is likely that inbreeding levels have become very high. However things are changing. In the Western world there is far less inbreeding now since people travel more and thus choose their marital partners from further afield. Britain is one of the most open societies in the world for marital choice. Historically marital partners speak the same language, have the same faith, and are the same race. Nowadays in Britain the most influential factor of who you will marry is educational level.

A study of surnames with an incidence of over 1% of the population in Britain shows how populations are dispersing. For example in 1881 the surname Jones was confined to west Wales but by 1998 it had spread into west England. A company called '23 and me' uses ancestry painting which looks at the source of the DNA on chromosomes to give information on a person's ancestral heritage. Many African-Americans use this service to find out their ancestral heritage. A Typically, an African–American woman has DNA of 65% of African origin, 29% European and 7% Asian, showing a great deal of DNA mixing. In the future, the source of DNA will become more and more scrambled as populations and people disperse.

To summarise, at least in the Western world (which is much of the world population). Evolution is not over but is going very slowly. So there is no need to worry about what utopia will be like - we are living it now!

References

1. Galton, F.: Hereditary Genius: An Inquiry into Its Laws and Consequences. Macmillan, London (1892)
2. Satoh, C., Takahashi, N., Asakawa, J., Kodaira, M., Kuick, R., Hanash, S.M., Neel, J.V.: Genetic analysis of children of atomic bomb survivors. Environ. Health Perspect. **104**(Suppl 3), 511–519 (1996)
3. Tiemann-Boege, I., Navidi, W., Grewal, R., Cohn, D., Eskenazi, B., Wyrobek, A.J., Arnheim, N.: The observed human sperm mutation frequency cannot explain the achondroplasia paternal age effect. Proc. Natl. Acad. Sci. U.S.A. **99**, 14952–14957 (2002)
4. Crow, J.F.: The origins, patterns and implications of human spontaneous mutation. Nat. Rev. Genet. **1**, 40–47
5. Tuljapurkar, S.D., Puleston, C.O., Gurven, M.D.: Why men matter: mating patterns drive evolution of human lifespan. PLoS ONE **2**, e785 (2007)
6. Lamason, R.L., Mohideen, M.A., Mest, J.R., Wong, A.C., Norton, H.L., Aros, M.C., Jurynec, M.J., Mao, X., Humphreville, V.R., Humbert, J.E., Sinha, S., Moore, J.L., Jagadeeswaran, P., Zhao, W., Ning, G., Makalowska, I., McKeigue, P.M., O'Donnell, D., Kittles, R., Parra, E.J., Mangini, N.J., Grunwald, D.J., Shriver, M.D., Canfield, V.A., Cheng, K.C.: SLC24A5, a putative cation exchanger, affects pigmentation in zebrafish and humans. Science **310**, 1782–1786 (2005)
7. Harris, S.S., Dawson-Hughes, B.: Seasonal changes in plasma 25-hydroxyvitamin D concentrations of young American black and white women. Am. J. Clin. Nutr. **67**, 1232–1236 (1998)
8. Robins, A.H.: Biological Perspectives on Human Pigmentation. Cambridge University Press, Cambridge (1991)
9. Darwin, C.: On the Origin of Species by Means of Natural Selection, or the Preservation of Favoured Races in the Struggle for Life. John Murray, London (1859)

Note: The editors have made every effort to trace the sources of the figures used in this chapter. If any sources are missing, due to unintentional oversights, the editors will be happy to add this information to later editions of this volume.

Evolutionary Medicine

Michael Ruse

Introduction

In this chapter I will discuss an ambitious, would-be entry into the field of health-care, namely "evolutionary medicine," the project of putting the whole of our understanding of and behavior towards those in need of attention on a firm evolutionary basis. Very much the brain child of two men, George Williams, the major twentieth-century evolutionist, and Randolph Nesse, a University of Michigan based psychiatrist, evolutionary medicine aims to revolutionize the field [1, 2]. As Dobzhansky [3] used to say, "nothing in biology makes sense except in the light of evolution," so Williams and Nesse would add "and nothing in medicine either makes sense except in the light of evolution."

Even though by 1959 (a convenient date being the hundredth anniversary of the *Origin*) neo-Darwinism – the synthesis of Darwinian natural selection theory and Mendelian (and later molecular) genetics – was an up and running paradigm, it was many years (if indeed in places barely now) before evolutionary theory was properly integrated into the biology undergraduate curriculum. In a way, Dobzhansky's statement was less a proud affirmation and more a plea for recognition. So likewise there has not exactly been a rush by health-care professionals (and more pertinently their teachers) to embrace the proffered insights of evolutionary theory. However, rather than bewailing (or celebrating) this fact, let us move directly to consider the claims made in a major recent textbook, co-authored by the eminent New Zealand scientist and physician, Sir Peter Gluckman. Let us follow him through an eight-fold classification of the ways in which evolution can impinge on the question of disease and our risk of suffering [4]. I will leave any philosophical reflections until we have finished.

M. Ruse (✉)
Department of Philosophy, Florida State University, Tallahassee, FL, USA
e-mail: mruse@fsu.edu

M. Brinkworth and F. Weinert (eds.), *Evolution 2.0*, The Frontiers Collection,
DOI 10.1007/978-3-642-20496-8_13,

What Is Evolutionary Medicine?

First, there is the fact that we (some individuals that is) might be in a situation for which evolution has not prepared us. Our environment or our culture has outstripped our biology. An example would be the oft-mentioned lactose intolerance. Since the coming of agriculture, about 8,000 years ago, there has been intense selective pressure on agriculturalists towards lactose tolerance that is to the ability through life to tolerate milk (and milk solids) from our domestic animals. Obviously most Europeans have now this acquired ability – although not everyone, and a recently fascinating suggestion is that it may have been at the root of Charles Darwin's ongoing ill-health [5] – but many in other parts of the world, without histories of agriculture, people do not. This can cause severe discomfort. Among other items Gluckman and co-authors mention is our inability to synthesize vitamin C, something that led to scurvy on board ships, until navies realized that daily drinks of citrus juice could avert the problem. Also there is obesity, perhaps not itself an illness but obviously one leading to such. It could be caused by the built-in desire to gorge when possible, something perhaps of great value in the Pleistocene but obviously much less so now in modern society.

Second are life-history associated factors. Most obvious here are the ailments of old age. Natural selection cares about getting us up to prime breeding condition and then keeping us fit so long as we are actively involved in child care and rearing. After that, we are on our own, and a very lonely "on our own" it can be too. We are much less able to handle infections and traumas and also have all of the diseases of old age – diseases that we would have evaded (because we would already be dead) in earlier times. Sometimes there is a direct connection between things that are useful earlier in life but harmful later on. Stem cells in tissues are a good example. While we are growing and reproducing, stem cells in tissues are of value because they promote tissue maintenance and repair. Unfortunately, later in life, they can lead to neoplasia, the abnormal proliferation of cells, that may perhaps be malignant.

Third there are excessive and uncontrolled defense mechanisms. Things like coughing, vomiting, and diarrhea are unpleasant in themselves, but they are fairly obvious ways in which a body tries to expel or ward off invading organisms. Obviously this can backfire if the mechanisms go into overdrive – dehydration following severe gastroenteritis is a case in point. Knowledge of the evolutionary significance of the mechanisms can be important in treatment. It is a commonplace that when we are sick we are often (usually) lethargic, disinclined to do anything very strenuous. There is evidence that this lethargy is part of our biology, slowing us down so the body can concentrate on fighting the sickness. Exercise when sick can be counterproductive. Fevers often fall in the same category. The usual advice is to take two aspirin and go to bed. But fevers are thought to be significant in fighting infection. Perhaps less so because they kill off bacteria directly and more so because they initial the production of certain proteins ("heat shock proteins") that can circulate in the blood and that have powerful anti-bacterial functions.

Fourth comes "losing the evolutionary arms race against other species." This is a very well-known and attested phenomenon. As soon as some new drug combating infection comes on the market, the organism or organisms against which it is directed are under huge selective pressure to develop ways to respond. Given the high rate at which organisms like bacteria reproduce – and their numbers – there is little surprise that it is but a few years before resistant strains are known. This occurred in the case of penicillin, introduced around 1942 and invaluable during the War. Within a year of the War's end, resistant strains were emerging. What makes this all a major problem is that nowhere do such resistant organisms emerge more quickly and strongly than in hospitals, the very places where people need most to be protected. This stems from a number of reasons, including the number of people who are sick, the high use of antibiotics, the ways in which staff can transmit diseases, and more. Expectedly, the fight against disease is complicated by the fact that resistant organisms resist in many different fashions. There is no one way in which they evade drugs, a way that could be explored in the hope of finding a one-step solution for everything.

Fifth there is the matter of design or evolutionary constraints. Some things highlighted by Gluckman [4] are perhaps less constraints and more things left over by the evolutionary process. The appendix is a case in point. For us, appendicitis can be fatal and yet the appendix has little or no function. It is a throwback to the days when our ancestors ate huge and probably near-exclusive amounts of herbage and needed the appendix for digestion. More obviously a constraint bringing about compromise is that determining the size of the human brain at birth. The larger the brain the sooner the child will grow up to full size. However, the larger the brain the more danger to the mother, who has a birth canal determined in size by the demands on the pelvis for upright walking. Another constraint is the lower back, which has loads and stresses upon it thanks to upright walking, quite beyond anything experienced by the apes. And as another example, it may be that male breast cancer (about 1% of all cases) comes about simply because there is just no easy way of getting rid of the genes which, when properly primed, cause functioning breasts in females.

Sixth is disease due to the direct effects of natural selection as it "balances" good effects against bad. The classic case here is that of sickle-cell anemia. In parts of Africa, malaria is a dreadful threat, killing sizable proportions of the population. Any gene therefore that causes protection against malaria is going to be under strong positive selection pressure. It turns out that there is a gene that offers protection, but the catch is that it offers such protection only if it is present in a single dose – more formally, if it is heterozygous to the normal or wild type allele (gene). It affects red blood cells in such a way that if the body is invaded by malaria then infected red blood cells collapse and are removed by the white blood cells. Unfortunately, two doses (the sickling allele is homozygous) cause the red blood cells to collapse into a tell-tale crescent or sickle shape, and the carrier generally dies young of anemia. It is a very simple piece of mathematics to show that in a population the devastatingly ill effects of two sickle cell alleles is balanced by the good effects (the malarial protection) of one sickle cell allele – and this persists for generation after generation, unless something external disrupts things. Another possible case of "balanced heterozygote fitness" might involve cystic fibrosis. It is thought that it might be caused by genes that in lesser amounts protect against typhoid or tuberculosis.

Seventh we have sexual selection and its effects. There are good reasons why fighter pilots tend not to be old men and women of 60. Thanks to sexual selection, it is young males who are more prepared to take risks and to do dangerous things. It is they who have the right hormones pumping through their bodies, making them ready to fight and to compete, directly or indirectly for females. Of course, young male humans are not just sexual aggressors. We have all been selected for sociality, the ability to live in groups. This requires moderation and tempering other desires. So at the very least we may have psychological conflicts and at worse violence and the injuries to which this and other forms of risk taking can lead. Whether, as seems fashionable today for film stars and sports celebrities, extreme sexual desire and behavior should be labeled a sickness, is an exercise left for the reader.

Finally, eighth, we have the "outcomes of demographic history." We have already touched on these issues earlier in this chapter and before. Gluckman and his fellow authors are referring to the asymmetries that we find in groups with respect to various genetically caused illnesses because of evolutionary history. Ashkenazi Jews and Tay-Sachs disease is a case in point. There is not something special about being Jewish that makes for Tay-Sachs disease susceptibility. It is rather that by an accident of history the mutation got into the population and because breeding (up to now, at least) has tended not to occur across population borders, Ashkenazi Jews are especially susceptible. In this case, it is social factors primarily that have set up the barriers. In fact, these barriers have been greatly dismantled in the USA, where increasingly there are unions between Jews and Gentiles. In other cases, the barriers have been more physical or geographical. This may be combined also with bottlenecks when population numbers were greatly reduced before expanding again. The inhabitants of Finland may exhibit all of these things. They came in small numbers across the Baltic from Southern Europe and once settled were much isolated by geography and climate. Expectedly, they show patterns of illness that distinguish them from others. For instance, comparatively there is a low incidence of Huntingdon's chorea, of cystic fibrosis, and of PKU. There is a high incidence of type 2 diabetes and cardiovascular disease. None of this is to exclude the possibility of environmental factors. Finland is very different from Italy in both winter and summer. It is to say that evolutionary biology may have been very important.

Presuppositions

Move on now to ask about matters of possible philosophical interest. Obviously these are there and some are very obvious. We all know that abortion and sterilization, especially enforced sterilization, are highly contested moral issues. With the

Catholic Church taking strong stands against both, there is little surprise to find that it was one of the leaders against eugenics, especially of the negative kind which wanted to restrict the reproduction and the production of the unfit. One suspects that with the growth in influence of the Catholic Church in the USA (not the least on the Supreme Court), any eugenic proposals like those floated (and often enacted) in the USA in the earlier parts of the twentieth century would have a much rougher ride today. So obviously there are philosophical issues here, although perhaps they are more general issues that are more exacerbated and posed by evolutionary medicine (broadly conceived to include such things as genetic counseling) than uniquely formulated by evolutionary medicine.

It is obvious also that there are going to be important conceptual and epistemological issues at stake. As one might expect of a field that was kickstarted by George Williams, today's evolutionary medicine tends to be hard-line Darwinian and individual selectionist at that. This does not mean that everything is thought to be an adaptation. No Darwinian, certainly not Charles Darwin, has ever made that claim. We have seen above that evolutionary medicine supposes that there are all sorts of places where things can get out of adaptive focus. There may be a lag between what was adaptive in the past and what is adaptive now. What is adaptive for one organism (a parasite) may not be particularly adaptive for another organism (us). Constraints and compromises and (what Darwin called) vestigial organs are another set of places where adaptation does not rule untroubled. Then there is the fact that sexual selection might be adaptive one way, but clearly might be highly counter-adaptive another way. And the final item, where we look at the effects of history on groups, shows that random factors – the founder effect especially – may play a crucial role in human health and disease. All of this, it goes without saying, is pretty standard Darwinian theory and has been stressed again and again in the century and a half since the *Origin*. The important point is that no new theory is demanded by the entry into the world of medicine.

None of this is to deny that the touchstone, the expected norm, is adaptive advantage. The whole point of evolutionary medicine is that we are looking at the body as a product of natural selection, and we expect to see adaptive advantage. Obviously in many cases we see this straight off. Eyes are for seeing and blindness is an affliction. Noses are for smelling although as Gluckman notes our evolutionary history rather points to the obvious fact that we rely a lot less on smell than do other mammals like dogs and hence the sense of smell is nothing like as crucial as the sense of sight. We have special schools for children who are blind. We have special schools for children who are deaf. We do not have special schools for children who lack a sense of smell. Although perhaps in the Pleistocene we might have needed them, because then a sense of smell might have been more vital – sniffing out meat that has gone off, for instance.

The commitment to individual selection is absolutely crucial in some instances where it is thought that evolutionary medicine has made triumphant breakthroughs in understanding. Harvard biologist Haig [6], for instance, has studied the relationships between mothers and offspring. You might think that here at least we are going to get one big happy family, but Haig – drawing on earlier individual-selectionist thinking by one of the founders of sociobiology Trivers [7] – notes that we may well get "parent-offspring conflict." What is in the interests of a mother may not be in the interests of a child and vice versa. Really, this is obvious as soon as you think about it. If a mother has two children it may be in her interests to split her attention between them or perhaps to give the younger child more attention. It does not follow that this is in the biological interests of the children, particularly the older child, even after we have factored in the relationship between the children (especially if there are different fathers). Haig applies this thinking to the circulation of the mother's blood. It is in the interest of the fetus to raise the level of proportion that goes to it; it is in the interest of the mother to moderate the circulation of the blood that goes to the fetus. The way this is played out is by the resistance that can be set up to the mother's circulation, and that is a function of her blood pressure. Blood pressure drops early in pregnancy and Haig's claim is that this represents the triumph of mother over fetus. Then later it starts to rise as the fetus now directs more blood in its direction – from virtually nothing at the beginning of the pregnancy to 16% at the end of the pregnancy.

Of course in a way you might rightly argue that this all functions in such a way that the fetus does well but the mother survives – they both have those interests in common. But sometimes in pregnancy women develop preeclampsia, a very dangerous condition associated with very high blood pressure, together with lots of protein in the urine. The obvious interpretation is simply that something has gone wrong. Haig [6] suggests rather that it might be a move of desperation on the part of the fetus. If for some reason it is not getting sufficient nutrients, it is in its interests to up the mother's resistance to that blood flow benefiting her (the mother) alone and the way to do this is by taking the blood pressure up even to dangerous levels. Interestingly preeclampsia is more common with twin pregnancies, and this of course is precisely a case where one individual fetus might not be getting enough. The fetus is gambling that it might be better off taking what it can now, even though it runs the risk of losing out on care later. The mother's interests do not enter into the equation, or only secondarily. About as individual-selectionist perspective as you could get.

The point being made here is obviously not that one now has a solution to preeclampsia or that one should refrain from interfering because this is "nature's way" or some such thing. At most, the point is that we may now have some true insights and that knowledge is the beginning of successful action. Also the point is not that individual selection theory is right and all-conquering. Rather that this seems to be the general pattern in evolutionary medical explanations today and needs to be recognized by those who would propose alternative explanations. Clearly there is need of conceptual analysis, for already some working in the field have appropriated terms like "multi-level selection." A case in point comes in a recent discussion by Bergstrom and Feldgarden [8] of ways in which one might apply insights from evolution to the creation of new barriers to invasive organisms. They point out that the dangers posed by bacteria are often not from the individual bacteria as such, but rather when they are in groups and start acting together. In other words, when we have a "quorum." Perhaps therefore a solution might be in tricking the bacteria into thinking (highly metaphorical language here!) that there is no such quorum. Moreover, and very desirably, when the bacteria social behavior is disrupted, it might not rebound as quickly as one might suppose. They write (referring to the ideas of others):

Where bacterial cooperation occurs, it is not an unavoidable consequence of direct individual selection as antibiotic resistance usually is, but rather a finely balanced consequence of multilevel selection. Thus if bacterial cooperation is disrupted, it may not return as readily as individually selected traits. To see how this might work, imagine a population of bacteria in which social behavior has been halted by disrupting quorum sensing. Whereas with conventional antibiotics the first antibiotic-resistant mutant has a substantial growth advantage, with quorum-sensing disrupters the first resistant mutant has a growth disadvantage. It provides a public good by producing constitutively, but it receives no benefits from the other members of the population who are not producing due to the quorum sensing disrupter. Moreover, because these behaviors are selected at the population level, if resistance does evolve it is likely to do so on the time scale of populations, rather than on the time scale of individuals. While a bacterium may reproduce in a matter of hours, populations often turn over on scales of weeks to months and thus resistance to quorum sensing disruptors is likely to evolve much more slowly than does resistance to conventional antibiotics [8].

Reading this the first time leaves a clear impression that group selection is at work – "behaviors are selected at a population level." However, if you look carefully, no such mechanism is really being proposed. The behaviors occur at the population level, but because they do not at first serve the interests of individuals – "it receives no benefits from the other members of the population" – they do not spread quickly. Indeed, one might even ask why they spread at all. "Multilevel selection" is a term being used not to incorporate group selection, but to acknowledge that individual selection can have group effects that are going to be important to the individual.

Sickness and Health

Apart from these issues to do with the kind of evolutionary theory being proposed or rather presupposed by those working in the field of evolutionary medicine, there are clearly other topics of great philosophical interest. For instance, very obviously there are issues to do with testing and how one might check out theories that apply to human beings. One cannot simply run experiments as one might on mice or rabbits. However, I want now to turn to a topic that lies behind any philosophical inquiry into the nature of medicine. I refer to the concepts of sickness, disease, and health. In recent years, much has been written on these topics, and it is important to see how they play out in the context of evolutionary medicine. This discussion will be a two-way process. What have the philosophers' discussions to say about evolutionary medicine and what has evolutionary medicine to say about the philosophers' discussions?

Leaving for a moment the question of health (and whether it is just a reverse of sickness), there are two basic approaches to the key problem of disease. These usually go under the headings of "naturalism" and "normativism," although other terms have been proposed [9, 10]. Philip Kitcher for instance proposes "objectivism" and "subjectivism."

Some scholars, objectivists about disease, think that there are facts about the human body on which the notion of disease is founded, and that those with a clear grasp of those facts would have no trouble drawing lines, even in the challenging cases. Their opponents, constructivists about disease, maintain that this is an illusion, that the disputed cases reveal how the values of different social groups conflict, rather than exposing any ignorance of facts, and that agreement is sometimes even produced because of universal acceptance of a system of values [10].

Whatever the language, you can see that the key divide is between those who think that disease is something "out there," that can be defined in terms of actual physical facts, and those who think that disease is necessarily a value notion, and as such is a matter of subjective or cultural ideas or themes or preferences.

Start with the naturalist position. The standard account is provided by Boorse [11–13]. He states flatly that: "On our view, disease judgments are value neutral... their recognition is a matter of natural science, not evaluative decision" [12, p. 543]. But how does one cash out the reference to natural science? In some sense, it has to be a matter of what is normal or natural for the species. "There is a definite standard of normality inherent in the structure and effective functioning of each species or organism ... Human beings are to be considered normal if they possess the full number of ... capacities natural to the human race, and if these ... are so balanced and inter-related that they function together effectively and harmoniously" (p. 554). But how are we to articulate the "definite standard of normality"? There's the rub! Suppose you just work statistically, and argue that the standard is the majority is the norm. Does this mean that being a minority in itself makes you sick? Apart from the tricky issue of sexual orientation – is a homosexual sick simply because he or she is in the minority? – think of the case of sickle-cell anemia. We certainly want to say that if the notion of disease comes in anywhere, it comes in here. But it is far from obvious that we are making this judgment simply because the sufferers are in the minority. We are making it because they are in desperate pain and will die young.

Perhaps therefore we should think more in terms of effective and harmonious functioning. Ignoring the group-selection hints in the above definition, presumably what we are now thinking of is to be cashed out in evolutionary terms, that is to say survival and reproduction. In a way, that seems to be pretty good, and attractive from the viewpoint of evolutionary medicine. If someone has a childhood leukemia, they are diseased because their prospects of survival and reproduction are much reduced. Similarly, if someone is losing out in an arms race with bacteria, again prospects of survival and reproduction are grim. However, we obviously run into problems very quickly. Suppose someone is vomiting and has diarrhea and a high temperature. Evolutionary medicine says that this is the body's way of kicking in and combating an infection. Do we want to say that such a person is not sick? "Pull your socks up and don't whine!" Sickle-cell anemia makes the situation even worse. Here we have something positively promoted by natural selection, at least in the sense that selection keeps the numbers up so that the heterozygotes do better in the struggle for survival and reproduction than they would otherwise. It is a very typological view of species – one that the late Mayr [14] spent a very long life-time combating – to say that a minority, being produced by natural selection, is not in some way typical of the species and even in some way part of natural functioning. As Randolph Nesse is always saying, you have got to stop thinking of natural selection as promoting health and happiness (whatever these might be). It is in the survival and reproduction business completely and utterly [1].

Proximate Versus Final Causes

Perhaps at this point it would be useful to invoke the distinction between proximate and ultimate causes. Obviously an evolutionary perspective focuses on ultimate causes, or final causes in the traditional language. Perhaps we should be looking in medicine always or primarily at proximate causes.

Schaffner [15] has argued very convincingly that although medicine might use teleological talk in its attempts to develop a mechanistic picture of how humans work, the teleology is just heuristic. It can be completely dispensed with when the mechanistic explanation of a given organ or process is complete. Schaffner argues that as we learn more about the causal role a structure plays in the overall functioning of the organism, the need for teleological talk of any kind drops out and is superseded by the vocabulary of mechanistic explanation, and that evolutionary functional ascriptions are merely heuristic; they focus our attention on "entities that satisfy the secondary [i.e. mechanistic] sense of function and that it is important for us to know more about" [15, 16].

Prima facie, this is an attractive move to make. A person with a temperature and the trots is sick no matter what the reasons in the long run. Look at what is making life so very difficult right now and get on with the process of helping and healing. Likewise the child with an appalling anemia is sick, has a disease, no matter what the ultimate reasons for this. That its siblings are thriving is in a way irrelevant. We want to know what causes the anemia, meaning the proximate causes, and how to tackle it.

Note that this is a broad ranging conceptual argument. You might load it down with additional points, for instance about the possibility that a disease has no direct final cause. It is not an adaptation, but a failure in adaptation. Or it was never really an adaptation in the first place, but perhaps a byproduct or the result of a constraint – the sort of thing that Stephen Jay Gould called a spandrel [17]. However, we have seen above that the evolutionary medicine supporter has the resources to deal with these issues, because the form of Darwinism that is presupposed takes these issues into account. (Whether practitioners always take them as fully into account as they should is perhaps another matter.) The question is whether looking at final causes, thinking teleogically, is a mistake in the first place. At best it can be used as a tool for discovery, as a heuristic. Here the response of the evolutionary medicine supporter will be that, however you want to define terms like disease and health, if you are in the business of health care then you really must look at final causes, you must ask Darwinian questions about adaptation. These are not merely heuristic. They are fundamental, and crucial to formulating adequate treatment. Should you give someone a couple of aspirin to bring down the temperature, or should you tell them to take it easy and tough it out? What about cases of preeclasmia? Should we just be focusing on the mother, or should we recognize that this may be a cry for help from the fetus and in treating the situation try to see that the fetus's needs are also taken into account? Should we see that evolution might tell us a lot about the fetus's needs? When faced with a difficult childbirth, should we recognize that "natural" may not necessarily be the best thing? We are faced with a compromise and natural selection has not been able to perform miracles. Hence, intervention in the form of caesarian births or at the least episiotomies are not to be proscribed in the cause of some false beliefs about naturalness or some such thing.

What about the actual use of language? Notwithstanding the significance of finding the evolutionary causes behind the phenomenon, it is hard to see how under any circumstances one would not want to speak of sickle-cell anemia as a disease. But perhaps in other cases one would want to modify the language. Perhaps a lot depends here on how revisionist one is going be about language use, or whether one is going to be conservative about these things. This is not a totally insignificant matter or mere question of taste. For instance, the medical definition of "obesity" today encompasses what in the past might have been described as "pleasingly plump," clearly a move made with the hope that those who are overweight even if not grossly so will be shocked into taking some remedial measures. In the same vein, namely of improving health care through the revision of language, one could possibly see a case for distinguishing cases where natural selection is working for the evolutionary ends of the individual (no matter how unpleasant) as opposed to cases where natural selection is working for the ends of others (healthy siblings, healthy babies) and cases where natural selection is simply failing (losing out in an arms race). Perhaps already, assuaging some of the worries of the linguistic conservative, we do some of this. Knowing the true state of affairs, you might want to say that the disease is the bacterial infection that your body is fighting. Having a high temperature is an unpleasant side-effect but certainly not a disease in itself or even part of the disease proper – whereas perhaps a swollen organ brought on by the bacteria is part of the disease proper.

Values

Obviously however all of this discussion is a bit truncated and distorted because we are not bringing in something that even the naturalist must take account of in some way, namely values. Why do we want to say that sickle-cell anemia is a disease? Ultimately, clearly, because it is unpleasant. People with sickle-cell anemia hurt.

Obviously at some level this is what lies behind Schaffner's urging us to think in terms of proximate causes. It is at the proximate level that pain and suffering occur. Final causes may be useful to understanding, but basically by definition they are not dealing with the here and now, and it is this that counts ultimately in medicine. As it happens, Boorse himself acknowledges this issue, for he makes a distinction between disease and sickness or being ill. It is the second term that carries the burden (if such it be) of values. Being ill is having a disease that we do not want, because it is unpleasant. Notice that values alone will not do the trick. My body must be broken down in some way biologically. I am in jail, awaiting execution for a crime that I did not commit. Undoubtedly I will be very sad and lengthy time on death row might make me clinically mad. But my essential sadness is no illness but a natural reaction to misfortune. (I specify that I did not commit the crime because with many people who did commit crimes their mental health is already in question.)

Of course introducing values does not solve the epistemological problems about defining disease purely naturally. Perhaps therefore the time has come to move right over to a normativist account of disease, what Kitcher calls constructivist. H. Tristram Engelhardt Jr. [18, p. 259] is the point person here: "We identify illnesses by virtue of our experience of them as physically or psychologically disagreeable, distasteful, unpleasant, deforming". Of course this is not enough. I am not ill if I am unjustly condemned to death. At once the normativists start moving over towards the naturalist side. We identify them "by virtue of some form of suffering or pathos due to the malfunctioning of our bodies or our minds. We identify disease states as constellations of observables, related through a disease explanation of a state of being ill" [18, p. 259]. Note that Engelhardt does seem to be assuming some kind of evolutionary, teleological understanding of causation. Others in the normativist camp seem more wedded to just proximate causes. Reznek [19] for instance wants to get away from the notion of malfunctioning and talk more in terms of abnormal processes. A major point here is that, once you give up the supposedly objective science as your measure of disease and move to something like abnormality, not only are you moving to proximate causes but you are moving into the realm of culture where what counts as an abnormality itself requires a value judgment. In other words, medical problems are what medical people deal with! A little less tautologically, what is to count as a disease requires a value judgment in itself. For instance, in many societies the desire of a man to have sex with as many women as possible is considered normal if sometimes social awkward; in America, apparently, it is an ailment calling for treatment.

Obviously, the exponent of evolutionary medicine cannot accept this at all. There can be agreement that pain and suffering is important in judging whether or not someone is sick, and that this might go back to the question of whether or not someone has a disease. There might be some sympathy for defining diseases in proximate cause terms, so long as this in no way denies or leads away from the essential importance of thinking in terms of adaptation, final causes. But ultimately whether or not there is something medically wrong cannot be a judgment from culture. Culture might be important in the judgment. The first of Gluckman's categories deals with ailments that come from rapid changes in environment, and culture is a key factor here. But the judgment is not itself cultural. In the end, it all comes down to survival and reproduction. Some do, some don't. It is as simple (or complex) as that.

Health

What about the flip side, what about health? To repeat Nesse again, natural selection does not care about you being healthy. It cares about survival and reproduction. Whether or not it is a disease, suppose your sexual obsessions are making you downright miserable. Instead of being able to settle down to a comfortable evening of reading the *Critique of Pure Reason*, you feel compelled once again to haunt the singles bars, engage in trivial and insincere conversation, all for the hope of a night of sex. If this is a better way of passing on your genes, then so be it. Unless it can be shown that your behavior backfires in some way, perhaps through the spread of STDs or perhaps through being able to provide proper parental care to those children who are born because of your behavior, it really doesn't seem that evolutionary medicine has a dog in this fight.

Obviously this is a little bit extreme. Are there more subtle ways in which biology might be connected to health? Some people want to define health in a way that refers to oneself essentially. Others like the German philosopher Gadamer [20], put things more in a social context: "it is a condition of being involved, of being in the world, of being together with one's fellow human beings, of active and rewarding engagement in one's everyday tasks" [20, p. 113]. And some want to combine the two. The World Health Organization defines health as "a state of complete physical, mental and social well-being and not merely the absence of disease or infirmity" [21]. However one decides, as the WHO definition makes clear, probably one is not going to define health purely in terms of survival and reproduction. Having a sense of fulfillment and being worthwhile is part of being healthy, and obsession with numbers of children is surely odd to the point of imbalance somewhere. Nevertheless, having children may be a very important part of what one considers full and healthy living. There are those who regard DINKS by choice (Double income, no kids) as if not sick then sadly truncated as human beings. Moreover unless you are essentially free from disease and handicap you are probably less likely to have total fulfillment and thus are less likely to be judged totally healthy. So biology surely does come in somewhere and the pertinence of the evolutionary approach is not to be denied totally or even in large part.

Exactly how this is all to be worked out is obviously a task for the future, and one suspects that philosophers could have much of worth to contribute. And this reflection, put in a broader context, is a good point on which to end our discussion. For all of its historical antecedents, evolutionary medicine as a formal approach is a relatively new discipline, perhaps with much promise but with far to go, both as science and medicine and as something part of medical organization and (very importantly) teaching. It raises some philosophical issues of great interest and those trained in the field both could and should get involved. My hope is that this short introduction will stimulate others to pick up the torch and to carry on the inquiry.

References

1. Nesse, R.M., Williams, G.C.: Why We Get Sick: The New Science of Darwinian Medicine. Times Books, New York (1994)
2. Nesse, R.M., Williams, G.C.: Evolution and Healing: The New Science of Darwinian Medicine. Weidenfeld & Nicholson, London (1995)
3. Dobzhansky, T.: Nothing in biology makes sense except in the light of evolution. Am. Biol. Teach. **35**, 125–129 (1973)
4. Gluckman, P., Beedle, A., Hanson, M.: Principles of Evolutionary Medicine. Oxford University Press, Oxford (2009)
5. Dixon, M., Radick, G.: Darwin in Ilkley. History Press, Stroud (2009)
6. Haig, D.: Intimate relations: evolutionary conflicts of pregnancy and childhood. In: Sterns, S.C., Koella, J. C. (eds.) Evolution in Health and Disease, 2nd edn., pp. 65–76. Oxford University Press, Oxford (2008)
7. Trivers, R.L.: Parent-offspring conflict. Am. Zool. **14**, 249–264 (1974)
8. Bergstrom, C.T., Feldgarden, M.: The ecology and evolution of antibiotic-resistant bacteria. In: Sterns, S. C., Koella, J. C. (eds.) Evolution in Health and Disease, 2nd edn., pp. 125–137. Oxford University Press, Oxford (2008)
9. Ruse, M.: Homosexuality: A Philosophical Inquiry. Blackwell, Oxford (1988)
10. Kitcher, P.: The Lives to Come: The Genetic Revolution and Human Possibilities, 2nd edn. Simon & Schuster, New York (1997)
11. Boorse, C.: On the distinction between disease and illness. Philos. Public Aff. **5**, 49–68 (1975)
12. Boorse, C.: Health as a theoretical concept. Philos. Sci. **44**, 542–573 (1977)
13. Boorse, C.: Concepts of health. In: C. Boorse (ed.) Health Care Ethics, pp. 359–393. Temple University Press, Philadelphia (1987)
14. Mayr, E.: Systematics and the Origin of Species. Columbia University Press, New York (1942)
15. Schaffner, K.: Discovery and Explanation in Biology and Medicine. University of Chicago Press, Chicago (1993)
16. Murphy, D.: Concepts of health and disease (Zalta, E.N. 2008)
17. Gould, S.J., Lewontin, R.C.: The spandrels of San Marco and the Panglossian paradigm: a critique of the adaptationist programme. Proc. R. Soc. Lond. B Biol. Sci. **205**, 581–598 (1979)
18. Engelhardt, H.T.: Ideology and etiology. J. Med. Philos. **1**, 256–268 (1976)
19. Reznek, L.: The Nature of Disease. Routledge, London (1987)
20. Gadamer, H.-G.: The Enigma of Health. Stanford University Press, Stanford (1996)
21. World Health Organization (WHO): WHO definition of health. Preamble to the Constitution of the World Health Organization as Adopted by the International Health Conference, New York (1946)

Note: This article is based on the final chapter of my forthcoming book, *Human Evolution: A Philosophical Introduction*, to be published by Cambridge University Press.

The Struggle for Life and the Conditions of Existence: Two Interpretations of Darwinian Evolution

D.M. Walsh

Among Darwin's many great achievements was the demonstration that the two central puzzles of the biological world—the distribution and adaptiveness of form—are consequences of a single process: evolution or 'descent with modification'. In the century and a half since Darwin set out his theory, it has been revised, refined and extended in scope, breadth and detail. While the history of the development of Darwinian thinking is one of expansion, it is also one of entrenchment. Darwinian thinking, as it has developed primarily through the twentieth century, has settled into a comfortable orthodoxy, commonly known as the 'Modern Synthesis Theory' of evolution. The Modern Synthesis is consistent with Darwin's theory, but it is also quite significantly an extension. The question has only just begun to be asked whether the twentieth century Modern Synthesis theory is the only reasonable possible extension.

In this essay I argue that it is not. Only recently, empirical advances in the understanding of organismal development, the inheritance of characters, and the mechanisms of adaptive change have begun to hint at an alternative. This alternative has yet to find a precise or wholly adequate articulation, but in its vaguest outline it is most evidently a starkly contrasting formulation of Darwinism. The principal difference between the Modern Synthesis rendition of Darwinism and its incipient competitor is the central explanatory role that the latter accords to the capacities of organisms, particularly as manifest in ontogeny, as the engine of adaptive evolution.[1] My strategy is to trace in outline the major conceptual developments leading from Darwin to the Modern Synthesis, and to suggest that these developments are not obligatory, especially given what we now know about development, inheritance and adaptation. I shall contrast the Modern Synthesis theory with this nascent organism-centred conception of evolution. While these two approaches are

[1]David Depew [1] calls this alternative 'Developmental Darwinism'. See also [2]

D.M. Walsh (✉)
University of Toronto, Toronto, ON, Canada
e-mail: denis.walsh@utoronto.ca

M. Brinkworth and F. Weinert (eds.), *Evolution 2.0*, The Frontiers Collection,
DOI 10.1007/978-3-642-20496-8_14, © Springer-Verlag Berlin Heidelberg 2012

radically different, they are both correctly to be seen as extensions of the theory of evolution adumbrated in Darwin's *Origin of Species*.

In order to be able to appreciate these as extensions of Darwin's account of evolution, we need to understand the core of the theory set out in the *Origin of Species*. I proceed in the following way. In sections "Darwin's Two Principles" and "Natural Selection", I present what I take to be the central insight of Darwin's theory. I argue that, according to Darwin, natural selection is not a cause of evolutionary change, but what I call a 'higher-order effect'. Darwin locates the causes of evolutionary change in the struggle for life in the conditions of existence. But it is unclear *how* the struggle for life in the conditions of existence manages to cause evolutionary change. The Modern Synthesis theory offers one compelling account (see section entitled "replicator Biology" below). But there is an alternative, organism-centered conception of the causes of evolutionary change that, I believe, is currently gaining empirical support. These two versions of evolutionary theory have significantly different commitments. They differ, crucially, on the canonical unit of evolutionary explanation—replicators or organisms. They further differ on the nature of the relation between the struggle for life and the conditions of existence (see "Organism-Centered Evolutionary Biology" below), and on the central question whether evolution is ineluctably chancy (see "Organism/Environment Relations" below).

My purpose in outlining these two alternative interpretations of evolution is simply to make the point that after a century and a half Darwin's theory of evolution continues to be fertile ground for evolutionary theorizing. It is as vital and relevant to current biology as it was to the biology of its day. Fecund as it is, Darwin's theory radically underdetermines the content of a complete theory of evolution. It is consistent with a variety of interpretations of the causes of evolutions. Here I present two drastically divergent ones.

Darwin's Two Principles

Darwin's theory of descent with modification is driven by two principles. The Struggle for Life and the Conditions of Existence. In Chapter three of the *Origin* Darwin asks his crucial question:

> How have all those exquisite adaptations of one part of the organisation to another part, and to the *conditions of life*, and of one distinct organic being to another being, been perfected? . . . [3 p. 114].

It is commonly supposed that his answer to the question is obviously 'natural selection' the process that Darwin discovered. But the very same paragraph ends with Darwin's real answer:

> All these results, as we shall more fully see in the next chapter, follow inevitably from the *struggle for life*. [3 p. 115].

The struggle for life comprises the complete suite of an organism's activities: its growth, nutrition, reproduction, the manner in which it exploits the resources of the environment. The struggle for life alone doesn't cause evolution, according to Darwin. At least the activities of organisms alone aren't sufficient to constitute struggle. As is well known, Darwin was greatly inspired by Malthus who argued that populations have a natural tendency to grow exponentially. If this propensity is left unchecked in the absence of 'struggle', Darwin realised, even the most slowly reproducing organisms could become so numerous as to cover the entire surface of the earth in a relatively short time. Something is needed to hold this natural tendency of populations in check. Only when this natural prodigality is constrained, there is struggle.

It is for this role that Darwin appropriated Lyell's conception of the economy of nature. For Lyell an ecological community constitutes a complex arrangement of organisms, each contributing to the stability of the system. It is an economy. Each species has a station in that economy: a role that it is uniquely suited to fill. Lyell used the concept of the economy of nature to explain the differences between fossil and recent faunas of the same region. As species become extinct, Lyell hypothesised, stations in the economy of nature are vacated. God then creates other species to fill those roles.

Darwin puts the economy of nature to different use. The economy of nature sets parameters on organisms' conditions of existence. The conditions of existence contribute both a limit and a filter for biological form. Those organisms more adept at negotiating the limits and conditions imposed on them by the economy of nature, survive and leave more offspring. Those variants that are better suited to that role are preserved in the struggle for existence. Darwin acknowledged that the economy of nature applies to parts of organisms too. The demands on an organism also place demands on the harmonious functioning of their parts. So the integration of the various parts of an organism into a functional whole can also be explained by appeal to the economy of nature. Darwin, unlike Lyell, held that the economy of nature is changeable. An organism's exploiting the resources of its environment, and competing with others can actually alter the environment [4].

The central point here is that Darwin's theory takes on certain metaphysical and explanatory commitments. Firstly, it is committed to a specific view on the nature of organisms: the salient feature of organisms is their struggle for life. Secondly, it is committed to the central place of organism/environment relations, as enshrined in the conditions of existence, as the cause of adaptation and diversification. These conjointly are the causes of evolution.

Natural Selection

Natural selection, it is routinely claimed, is Darwin's great discovery. Yet my account of Darwin's theory doesn't mention it. How can an account of the causes of evolution neglect to mention natural selection? It is absolutely clear that Darwin discovered a heretofore unknown process—natural selection. This process, in turn, issues in a distinctive type of explanation. But it doesn't follow that selection causes evolution. This is an important issue because taking natural selection to be a cause of evolution is a crucial first step in the transmogrification of Darwin's theory into its Modern Synthesis successor.

The Metaphysics of Selection

It is unclear what metaphysical role Darwin took selection to play. To be sure, Darwin sometimes appears to accord natural selection a distinctive causal role. Natural selection is spoken of as a mechanism. It is said to be 'daily and hourly scrutinising' etc. But this is clearly metaphorical speak. There is another possible interpretation, selection as what I shall call a 'higher-order effect'.

Darwin's express wish was that his theory of descent with modification should achieve at least the same success in explaining the fit and diversity of form as Paley's argument from design, although without an appeal to intentional agency. For Paley [5], the magnificent adaptedness of organisms to their conditions of existence and the staggering array of biological forms are the work of a beneficent designer. The appeal to a designer is predicated on the supposition that the characteristic activities of organisms alone are insufficient to explain the adaptedness and diversity of biological form. Something more, *an additional adaptation-promoting cause,* is needed. This is a compelling idea and it plays a very significant role in the development of evolutionary thought.

It is a thought that falls easily out of Darwin's own line of argument. Darwin motivates his claim for natural selection by pointing to the power of a non-natural analogue—artificial selection. Breeders can effect significant changes in the form of a species simply by picking variants and crossing them. In artificial selection there is a process over and above the normal processes occurring within individuals which is causally responsible for changes in form. This process is selective breeding. Darwin argues by analogy that what breeders could do in a few generations, selecting only a few characters, nature could do even more comprehensively.

> ...natural selection as we shall hereafter see is a power incessantly ready for action, and is as immeasurably superior to man's feeble efforts, as the works of Nature are to those of Art [3 p. 115].

Darwin's invocation of selective breeding is a stroke of rhetorical genius. Selective breeding occupies a point midway between the intentions of a designer and the natural processes occurring within individual organisms. Breeders are intentional agents, but they work through the medium of those processes endemic to organisms, mating and the inheritance of variant forms.

The standard interpretation of this rhetorical device is that for Darwin, just as artificial selection is a mechanism that operates over and above the activities of organisms so is natural selection. Certainly talk of natural selection as a cause or 'power' is prevalent throughout the *Origin*. Moreover, much has been made of Darwin's commitment to *vera causa* explanations. Darwin following Herschel thought that the proper object of scientific investigation was the discovery of true causes [6] He took pride in the thought that his theory conformed to this proper methodology. Accordingly, generations of Darwin commentators from Helmholtz [7] to Hodge [8] have interpreted Darwin as holding that natural selection is a cause, even a mechanism, of evolutionary change.

Indeed, this has become the standard interpretation of both Darwinian and modern evolutionary theory. Selection is seen in both to play the role of a mechanism, sometimes even a force. Elliott Sober's influential treatment exemplifies the point. Sober takes the concept of 'fitness' to stand for the propensity of individuals to survive and reproduce (in the struggle for life). According to Sober, fitness doesn't cause evolutionary change; selection does:

> Selection for is the causal concept par excellence... An organism's overall fitness does not cause it to live or die, but the fact that there is selection against [say] vulnerability to predators may do so. [9 p.100].

Like a breeder, selection causes the differential survival and reproduction of organisms.

An alternative, and to my mind more plausible, interpretation of natural selection is that it is not a causal process, even according to Darwin. It is simply what I shall call a 'higher-order effect'. It does not cause individuals to live, die or reproduce. It is not a further cause over and above the natural activities of organisms. It is an ensemble-level process, but not a causal process. Natural selection is an aggregate of causal processes: those processes that constitute the natural activities of organisms in the struggle for life—birth, death and reproduction.

Not all aggregates of causal processes are themselves causal processes. The movement of a shadow across the ground may be the aggregate of the movement of an object and the differential illuminations of the ground surface. This is a process; it is an aggregate of causal processes, but famously, not a causal process [10]. The Coriolis effect is the aggregate of the movement of a body along the surface of the earth and the movement of the earth. It too is a process, but not a causal process. Some such aggregates are even colloquially called 'forces', for example centrifugal force is the aggregate effect of the inertia of a moving body that tends to continue its rectilinear motion and the centripetal force drawing the body in toward the centre of motion. It is felt as an 'outward force', but of course it is not.

In a similar way, Walsh [11] has argued that natural selection is a pseudoprocess: it is a non-causal, ensemble-level aggregate of individual-level causal processes. Natural selection is the effect that the aggregate of births, deaths and reproduction has on the structure of a population. Matthen and Ariew [12], argue that the changes in the trait structure of a population, identified as selection, are simple 'analytic consequences' of the differential survival and reproduction of individuals. When organisms are born, die, or reproduce, the trait structure changes instantaneously. That change is known as 'natural selection'.

But if natural selection is not a cause of evolution, if it is merely an effect, how does it figure in genuine explanations of population change? In addressing this question, I believe, we encounter Darwin's real genius. Darwin not only discovered a new ensemble-level process—natural selection—he also discovered a whole new kind of explanation: I shall call it a 'higher-order effect explanation'.

Higher-Order Effect Explanations

Higher-Order effect explanations are actually quite common. Consider Erwin Schrodinger's [13] example of the explanation of passive diffusion. If we put a drop of Potassium Permanganate in a beaker of water, the effect we see is that the permanganate diffuses from the area of high concentration to areas of the volume with lower concentration, until, eventually, the permanganate is evenly distributed throughout the volume of water. How do we explain the directionality of the observed process? Schrodinger accounts for diffusion in the following way. Suppose we drop a membrane into the solution at some point before it has reached equilibrium. There will be more permanganate particles on one side of the membrane than the other. If we suppose that these particles are moving randomly, the frequency of permanganate particles colliding with the membrane will be greater on the high-density side than on the low- density side. We can take the membrane as offering a sample of the distribution of the directions of the particles' motion. There are more moving from high-density to low-density. So, the higher-order effect of aggregating the movements of the individual particles is that the system moves toward a state in which the permanganate particles are more or less evenly distributed throughout the solution.

This is an instance of what I am calling 'higher-order effect' explanations. It has some interesting diagnostic features. First off-it explains an ensemble level effect by citing the properties of members of the ensemble (not the properties of the ensemble). Tellingly, no ensemble-level force or cause is required; there is no 'diffusive' force. Diffusion is not a force or cause, it is simply an analytic consequence of the differential motion of particles. It consists of an ensemble-level bias—the tendency of the system to move toward an even distribution of particles. Yet this ensemble-level bias requires no individual-level bias. The motions of individual particles *ex hypothesi* is random. This unbiased motion at the individual level is enough to cause the biased motion at the ensemble level.

Higher-order effect explanations do not require that the lower-level processes they appeal to are unbiased. Another example from Schrodinger underscores the point. When we watch a fog descend, we see that it has a reasonably precise upper boundary and a constant rate of descent. Each of the individuals water droplets, however, is moving more or less randomly. Each has a very slight downward bias which may not be discernible by watching its trajectory. None of them moves at constant velocity or in a constant direction. Nevertheless the sum of these numerous tiny indiscernible biases is the discernibly constant and precise rate of descent of the fog as a whole. The higher-order effect arises from the aggregation of the slightly biased motions of the individual water droplets.

We can now see the importance of construing selection as a higher-order effect. Far from positing selection as an ensemble-level cause, Darwin's theory demonstrates that the activities that constitute the struggle for life—births, deaths, and reproduction of organisms—are sufficient by themselves to bring about changes in the trait structure of the population that constitute natural selection. So long as organisms struggle for existence, and there is variation in the population, descent with modification will occur. Evolution by natural selection, is in this sense, spontaneous. It doesn't need any further causes. Darwin's theory of the fit and diversity of form has the virtue of parsimony. Darwin demonstrates the falsity of Paley's metaphysical presupposition that we need to posit some cause or force *in addition to* the births and deaths of organisms. No causes other than organisms' struggle for life in the conditions of existence are needed to explain the fit and diversity of form.

Furthermore, it shows that the components of biological ensembles are explanatorily indispensable to Darwin's theory of evolution. The causes of evolution are not to be found at the level of population-level dynamics. They are to be found in what organisms do. This is radically in contrast with Elliott Sober's claim about the import of Darwin's population thinking. According to Sober [14 p. 370]: "population thinking is about ignoring individuals".

This raises the question 'how do the struggle for life and the conditions of existence conspire to cause adaptive evolution?'. There are (at least) two conceivable accounts. One of these has become the orthodoxy enshrined in the Modern Synthesis theory, the other is perhaps only now beginning to gain some currency. The principal difference resides in what each interpretation takes as the canonical unit of biological organisation: the replicator on one account and the organism on the other. There are three further corollary differences of particular significance. The two interpretations of the causes of evolution that I shall outline differ in the way that each construes the relation between the organism and its conditions of existence. They further differ on the appropriate mode of explanation required to account for adaptation, and they differ, on the role of chance in evolution. Despite the differences, each of these is an extension of the theory outlined in Darwin's Origin of Species.

Replicator Biology

The development of evolutionary theory throughout the twentieth century has seen the central role accorded by Darwin to the activities of organisms usurped by the activities of sub-organismal entities, replicators (genes). On the twentieth century embellishment of Darwinian evolution, organisms are no longer the canonical unit of biological organisation, replicators are. Replicators are sub-organismal entities that are copied and transmitted from parent to offspring. Genes are the paradigm example.

> Evolution is the external and visible manifestation of the survival of alternative replicators . . . Genes are replicators; organisms . . . are best not regarded as replicators; they are vehicles in which replicators travel about. Replicator selection is the process by which some replicators survive at the expense of others. ([15 p. 82)

If evolution is a higher-order effect of the activities of lower-level entities, then replicator biology is the conviction that the lower-level entities best suited to carry the explanatory burden are replicators. Replicators enjoy this particular privilege because of the distinctive contributions they make to the component processes of adaptive evolution: inheritance, development and adaptive population change.

Organismal evolution requires: inheritance, development and adaptive population change. Inheritance is required for the evolutionary change to be cumulative. Development is required for evolutionary change to be registered in changes of phenotype or form. Adaptive change is what secures the fit of organisms to their conditions of existence. These processes are different in character: inheritance is an inter-organismal process, the process that secures the resemblance of offspring to parent. Development is an intra-organismal process. Adaptation is a supra-organismal process—change in population structure.

Replicator biology incorporates two bold claims about these processes. The first is that they are distinct and quasi-autonomous. By this I mean that the process of development does not contribute to the process of inheritance, nor does it promote adaptive change in a population. Selection does that. Similarly the process of inheritance does not promote adaptation. Organisms inherit their parents' genotypes whether they are beneficial or not. Of course selection, does not change the content of the traits that are developed or inherited. It merely winnows. The only non-autonomy amongst this suite of processes is the asymmetrical dependence of development on inheritance. As far as evolution is concerned, with one significant exception—that of mutation—organisms only develop the traits they inherit. The second claim is that despite their being quasi-autonomous, there is a single unit of biological organisation that explains each process: the replicator.

Inheritance: Inheritance is the intergenerational stability of form. According to the Mendelian theory of inheritance that developed since Darwin's time, inheritance is particulate. Parents pass to their offspring replicated particles that encode the information required to build an organism. 'Inheritance' has come to stand for the process by which replicated material is copied and transmitted from parent to offspring.

Development: Development is the growth of an organism from zygote (in the case of sexually reproducing individuals) to adult. If the Mendelian theory accords replicators a special role in inheritance, then the doctrine originating with August Weismann confers on replicators a unique and special control over the development of phenotype. Weismann discovered that the material of inheritance is sequestered from the somatoplasm early in development (of metazoans at least). The germ plasm, the material in which replication occurs, is thus quarantined from any of the changes wrought on the somatoplasm during development. Changes made to the organism during development are not passed on to offspring. Only elements of the germ plasm are copies transmitted from generation to generation. In this way, replicated, inherited material not only plays a privileged role in inheritance, it plays a special role in development. Inherited material is said to encode information for building an organism.

Population Change: Because evolutionary biology involves only heritable change and only replicators are inherited, it appears to follow that replicators should have an especially important place in the modern synthesis account of evolutionary change. Sure enough, on this view, evolutionary change is measured and defined in terms of change in replicator structure in a population. Selection occurs through the differential survival and reproduction of organisms, yet the evolutionary effects of this differential survival and reproduction are caused by, measured as, and defined over, changes in replicator structure. Adaptive evolution is seen as the accretion and accumulation of advantageous replicators.

So, while evolution comprises three more or less discrete processes, there is a single canonical unit of biological organisation that unites and explains them. Replicators are (i) the units of inheritance, (ii) the units of phenotypic control and (iii) the units of evolutionary change: within a population

The development of twentieth century biology has seen the progressive displacement of the organism from its previous central place in the understanding of the fit and diversity of organic form. The organism is now a middle man in evolution—the interface between the replicating, organism building activities of genes and the selecting power of the environment.

> In this view the organism is the object of evolutionary forces, the passive nexus of independent external and internal forces, one generating "problems" at random with respect to the organism, the other generating "solutions" at random with respect to the environment [16] p. 47].

Organism-Centred Evolutionary Biology

There has been renewed interest in recent years in reviving the organism. Organisms are highly distinctive features of the natural world. They are self-building, self-organising, highly plastic, homeostatic, highly complex systems. In Evelyn Fox Keller's words, the organism is:

> a bounded physico-chemical body capable not only of self-regulation—self-steering—but also, and perhaps most important, of self-forming. An organism is a material entity that is transformed into a self-generating "self" by virtue of its peculiar and particular organization [17 p. 108].

At first blush it would seem rather unlikely that these features were not to make some important contribution to evolution. Yet, throughout much of the twentieth century, the emphasis has been on minimising the uniqueness of organisms [18, 19].

In recent years, however, one particularly distinctive capacity of organisms has been increasingly implicated in the causes of evolution [20, 21]. The general idea is that one of the essential features of organisms—plasticity—contributes not just to the capacity of organisms to succeed in the struggle for existence, it also contributes to the process of adaptive evolution. Plasticity is the capacity of an organism to attain and maintain a stable, viable form despite the vagaries of its conditions of existence (both internal and external) by making adaptive, compensatory changes:

> The organism is not robust because it is built in such a manner that it does not buckle under stress. Its robustness stems from a physiology that is adaptive. It stays the same, not because it cannot change but because it compensates for change around it. The secret of the phenotype is dynamic restoration ... [22 pp. 108–109].

This capacity of organisms to make compensatory changes to form or function in the face of the vicissitudes of genetics or environment is known as phenotypic plasticity. It contributes to the ability of organisms to succeed in the struggle for existence by conferring on them the capacity to mitigate the adverse effects of mutations or changes in environmental conditions.

But, according to an emerging opinion, it can do more as well [23]. Phenotypic plasticity contributes to adaptive evolution in the following way. Each part of an organism's developmental apparatus has a broad phenotypic repertoire. That is to say that each part has the capacity to produce a wide range of stable structures, including novel adaptive structures, under a range of conditions. The importance of phenotypic repertoire for the development of organisms cannot be overemphasised. Development requires an enormous amount of coordination and orchestration. If, for example, a muscle mass increases in response to some developmental demand, then concomitant effects on other systems are also required [24]. The structure of the surrounding bone must also change to accommodate the altered forces, so too must the amount of innervation and vascularisation [21]. Any adaptive change in one system is accompanied by adaptive, compensatory changes in other systems.

> An environmentally induced change in morphology, for instance, is often accompanied by changes in behavior and physiology. Hence, induction of one phenotype can indirectly influence the expression of numerous other traits and expose them to novel selective pressures. [24 p. 460–461]

West-Eberhard calls this coordination of developmental processes phenotypic accommodation. These accommodations promote the viability of the organism by minimising the disruption caused by the development of new forms [25].

Adaptive phenotypic change requires the orchestration of multiple developmental systems. Plasticity and phenotypic repertoire confer on an organism the capacity to ensure this adaptive orchestration of its various developmental sub-systems. If each of these sub-systems was under rigorous genetic control, that is to say if each of the concomitant phenotypic changes required its own genetic mutations, then the adaptive evolution of complex organisms might never occur. So, the phenotypic repertoire and the accommodation that is underwritten by the developmental plasticity of organisms is a necessary pre-requisite for complex adaptive evolution [20, 25].

Because development is so heavily buffered, in normal conditions it masks a significant amount of genetic and epigenetic variation. Adaptive phenotypic change exposes this underlying genetic diversity [23]. This diversity, in turn, is then available either for the production of new forms, or is co-opted for the reliable intergenerationally stable production of new forms, through a process of genetic assimilation:

> Genetic accommodation is a mechanism of evolution wherein a novel phenotype, generated either through a mutation or environmental change, is refined into an adaptive phenotype through quantitative genetic changes [24 p. 461].

Through genetic accommodation, novel adaptive phenotypes become routinised and entrenched. The developmental system enshrines those processes that most reliably produce the new adaptive novelties.

This model suggests an account of the causes of adaptive evolution that is radically divergent from the replicator model. On the organism-centered model, adaptive novelties arise in the development of organisms, and not by the mutation of replicators. Adaptive evolutionary change, furthermore, is initiated by the property that makes organisms the very things they are: highly plastic, self-building, self-regulating entities, capable of 'dynamic restoration'. This account of evolution reverses the priority of genetic change over phenotypic change. Adaptive phenotypic change occurs before genetic change. The adaptive responses of organisms to their conditions of existence causes change in the genetic structure of the population. "Genes are probably more often followers than leaders in evolutionary change" [26 p. 6,543]. Furthermore, the inheritance of novel forms is secured not just through genetic modifications, but through the regulation of organismal development. On this view, the adaptive plasticity of organisms is a pre-requisite for—and not merely a consequence of—change in the genetic structure of populations: "Without developmental plasticity, the bare genes and the impositions of the environment would have no effect and no importance for evolution. " [26 p. 6,544].

The principal difference between replicator evolutionary biology and organism-centered evolutionary biology is that whereas the former marginalises organisms' 'struggle for life' as the cause of evolutionary change, organism-centered biology prioritises—and even enriches—it. The struggle for life isn't merely the struggle against dearth. It is the purposive maintenance of viability against the vagaries of the internal and external conditions of existence. Organisms pursue the struggle for life through highly plastic, adaptive, 'dynamic restoration'. This, on the organism-centered approach, is the principal cause of evolutionary change.

Organism/Environment Relations

Replicator and Organism-Centered biologies differ not just one what they take to be the principal causes of evolutionary change. They also differ on how they conceive the relation between the organism and its environment. That is to say, not only do they differ on the importance of the struggle for life, they take radically different stances on the *conditions of existence*.

Replicator biology suggests a form of decoupling between organisms and their conditions of existence. Organisms are built by the activities of genes or replicators. Adaptive change comes about through the capacity of the external environment to alter biological form. The concept of the ecological niche plays a particularly important role here. Niches are very often conceived in the way that Lyell thought of the 'stations' in the economy of nature.' A niche, on this view, is a set of extra-organismal conditions that mould biological form to meet its exigencies. The decoupling of organism and niche is a necessary feature of the replicator approach to the explanation of adaptation. Replicator biology's adaptive explanations are externalist [27, 28]. They explain how the conditions of existence, conceived as the extra-organismal environment, causes the differential survival and reproduction of organisms, and with that, the differential loss and retention of replicators in a population. But the environment, the niche, can only fit form to meet its exigencies, if those exigencies are independent of, and prior to, adaptive form.

> To make the metaphor of adaptation work, environments or ecological niches must exist before the organisms that fill them. Otherwise environments couldn't *cause* organisms to fill those niches. The history of life is then the history of coming into being of new forms that fit more closely into these pre-existing niches [16 p. 63].

Organism-centered biology, by contrast, is, implicitly or explicitly, opposed to the decoupling the struggle for life from the conditions of existence. The relation, instead, is one of constructive interaction. Because development and the mechanisms of inheritance play a significant role in the process of adaptive evolution, they also play a significant role in determining the conditions of existence. When an organism makes an adaptive change to some feature of its phenotype, it alters the conditions of existence of other features of the phenotype. These features, as we have seen, are required to make compensatory, phenotypic accommodations, in turn. Moreover, these altered phenotypes constitute a new developmental context in which genes operate. Furthermore, as organisms make adaptive changes in response to their external conditions of existence, they alter the way those conditions affect survival and reproduction. Phenotypic adaptation changes the relationship of organism to environment in ways that buffer the organism against the deleterious features of the environment. In this respect, phenotypic adaptive change alters the way that organisms experience their environments. On this view, the conditions of the existence are best interpreted not as independent features of the extra-organismal environment, they are affordances provided by both internal and external features [29]. Adaptive change alters an organism's affordances:

> The affordances of the environment are what it offers the animal, what it provides or furnishes, for good or ill. ... I mean by it something that refers to both the environment and the animal .... It implies the complementarity of the animal and the environment [30 p.127].

Affordances, as I shall interpret them, can be either internal or external to organisms.

In adapting to internal and external conditions, organisms create a new set of affordances, to which they must further adapt. Organisms and the affordances to which they respond are thus mutually interactive. Organisms react to and at the same time alter the affordances that they encounter. That is to say, as organisms undergo adaptive phenotypic change in the struggle for life, they alter their conditions of existence. These new conditions, in turn, require further adaptive phenotypic changes. The Struggle for Life and the Conditions of Existence are thus not decoupled. They are entwined and mutually constructing. As David Depew [1] says: "[Organisms] are agents in making the very worlds to which, precisely because they make them, they are adapted".

This suggests, then, that rather than being externalist, adaptive explanation should be 'interactionist'. That is to say, rather than explaining adaptive evolutionary change by citing the influence of the external environment on genes, organism-centered biology should be in the business of explaining adaptive evolution by citing the plastic, adaptive interaction between organisms and the mutually constructing conditions of existence.

> through contributing to the environmental conditions of development for successor generations, organisms—including human beings—actively participate in their own evolution ([31 187).

There is a reciprocal, constructive interactive relation between organism and environment. The same reciprocal construction holds between an organism and its parts. Here too there is reciprocal adaptive interaction. In effect, the relation between an organism and its sub-systems is no different than the relation between an organism and its environment. They are both conditioned by the adaptive plasticity of organisms. This conception of the conditions of existence is strongly reminiscent of Cuvier's.

> Every organized individual forms an entire system of its own, all the parts of which mutually correspond, and concur to produce a certain definite purpose, by reciprocal reaction, or by combining towards the same end. Hence none of these separate parts can change in their forms without a corresponding change in the other parts of the same animal, and consequently, each of these parts, taken separately, indicates all other parts to which it has belonged. (Cuvier 1812 Discours Préliminaire. Quoted in [4])

Darwin's conception of the conditions of existence originates, of course, with Cuvier.

The important point here is that according to organism-centered evolutionary biology, the struggle for life, in effect, creates the conditions of existence, which in turn impact on the struggle for life.

This is all to illustrate that replicator biology and organism-centered biology conceive of the relevant conditions of existence very differently. On the replicator view, conditions of existence are wholly external to organisms. They are individuated by the extra-organismal physical features of the environment. They are not constituted by the activities of organisms. These external features alone are the agents of adaptive evolutionary change. On the organism-centered view, the conditions of existence are both internal and external to organisms. They are constituted by the plastic interaction of organisms with their physical setting.

Chance and Inherency

As a further corollary, replicator biology and organism-centered biology take radically different approaches to the explanation of adaptive evolution. The mode of explanation in replicator biology is exclusively mechanistic. We explain adaptive evolution by citing the mechanisms that bring about change. The causal activities of genes explain the inheritance and development of form. The causal impact of the environment on organisms explains the change in population structure. The active pursuit by organisms of the goals of survival and reproduction play no part in replicator theory. Consequently there is no role for the sort of purposive explanation that goal-directedness figures in [32].

One consequence of this exclusive reliance on mechanism is that evolution, for replicator theorists, is ineluctably chancy–'chance caught on the wing', in Jaques Monod's [33] evocative phrase. The ultimate source of the variation upon which selection acts is the random mutation of replicators. Organismal development and inheritance play no role in accounting for the adaptiveness of evolution, because they are seen as fundamentally conservative. Organisms inherit their replicators and develop the traits for which they code. There is nothing inherently 'creative' or adaptive about these processes. The source of the 'creativity' [34] or adaptiveness in evolution is ultimately the random production of variants, and the retentive power of selection:

> The initial elementary events which open the way to evolution in the intensely conservative systems called living beings are microscopic, fortuitous, and utterly without relation to whatever may be their effects upon teleonomic functioning ([33] p. 118).

Adaptive evolution depends fundamentally on non-biased random mutations. The higher-order directionality of the effect is built upon an underlying randomness, in much the way that the diffusion of potassium permanganate, discussed in our example above, depends upon the random motion of permanganate particles. New forms originally occur by chance, and then are locked in and built upon by selection. This is a deeply entrenched commitment of Modern Synthesis replicator biology.

> ...the non-random selection of randomly varying replicating entities by reason of their 'phenotypic' effects... is the only force I know that can, in principle, guide evolution in the direction of adaptive complexity [35 p. 32].

Perhaps the *locus classicus* of this commitment in the Modern Synthesis is to be found in Ernst Mayr's distinction between proximate and ultimate causes in evolution [36]. According to Mayr, development and inheritance are proximate causes of the possession of traits by individuals. But we cannot explain adaptive evolution by citing these causes. The cause of adaptive evolution, according to Mayr, is a process wholly distinct from development and inheritance, natural selection. To ask about the adaptive significance of a trait—'what is it for?', is to ask for its ultimate cause. Ultimate cause explanations advert to mutation and selection. Thanks to natural selection, Mayr argues, evolutionary explanation requires no unreduced appeal to the purposes of organisms—random mutation and selection suffice.[2]

Despite the welter of evidence that development is adaptive, many biologists and philosophers still maintain that it plays a role in explaining the evolution of form only insofar as it is essentially conservative [38]. It is worth pointing out that the conservativeness of form and the randomness of novel features is not a commitment of the theory adumbrated in the *Origin of Species*. Darwin readily accepts that the use and disuse of parts can be a cause of changes of form, and that such changes can be heritable. He offers evidence of this from breeding experiments.

> I think there can be little doubt that use in our domestic animals strengthens and enlarges certain parts, and disuse diminishes them; and that such modifications are inherited ([3] 175).

In contrast, organism-centered explanations of adaptive evolution do not, or should not, restrict themselves to pure mechanism. They can appeal to organisms' purposive pursuit of their goals as a contributing factor in evolution. Organisms are the very paradigms of goal-directed systems. Their characteristic activities—self-regulation, adaptive response, dynamic restoration—are *goal-directed* activities.

> You cannot even think of an organism ... without taking into account what variously and rather loosely is called adaptiveness, purposiveness, goal seeking and the like. [39 p. 45].

Phenotypic plasticity is the manifestation of this goal directedness. This goal-directed plasticity causes and explains adaptive evolution.

The general idea is that an organism's capacity to succeed in the struggle for life consists in its pursuit of goals. This pursuit of goals is manifested as adaptive changes to form and functioning throughout the organism's lifetime. Because organisms have the capacity to maintain and produce these novel adaptive phenotypes across generations, some of these changes are or become intergenerationally stable, that is to say, inheritable.

There is thus an extremely important set of regularities of the biological world that are not visible from the mechanistic perspective of random mutation and selection. Some changes to biological form occur *because they are conducive to the goals of organisms*. Organism-centred adaptive evolution occurs because organisms are capable of a particular kind of goal-directed activity. If this is correct, then from the perspective of organism-centered biology, adaptive evolution is not ineluctably chancy. But these teleological regularities are imperceptible from the perspective of replicator theory.

Newman and Muller introduce a distinction between what they call 'contingency' and 'inherency'.

> Something is contingent if its occurrence depends on the presence of unusual ... conditions that occur accidentally, conditions that involve a large component of chance, ... something is inherent either if it will always happen ... or if the potentiality for it always exists. ([40] p.)

[2]André Ariew [37] provides an engaging discussion.

Replicator and organism-centered theories take a very different position on the contingency or inherency of adaptive evolution. For replicator biology, adaptation depends ultimately on chance, or contingent occurrences, such as random mutation. Organism-centered biology takes the adaptiveness of evolution to be inherent in the goal-directed capacities of organisms as manifested in their phenotypic plasticity. A significant number of evolutionary novelties first appear, not by chance, but because they are conducive to the survival and reproduction of organisms. Adaptive evolutionary change is inherent in the purposive, goal-directed plasticity of organisms.

Conclusion

Darwin's *Origin of Species*, as we have seen, introduces two radical new theses. The first, of course, is that the fit and diversity of organismic form are the result of a single process: descent with modification, or evolution. The second insight, less often noticed but no less profound, is that evolution is a 'higher-order effect'. It is a process that occurs within an ensemble, that is caused and explained by the activities of the members of that ensemble. For Darwin, the relevant activities are the struggle for life that organisms undertake in the conditions of existence. But how does the struggle for life in the conditions of existence cause or explain the higher-order effect?

I have outlined two quite drastically different approaches to the understanding, and the study, of evolution. I have called these 'replicator' and 'organism-centered' approaches. Each is consistent with the theory of the fit and diversity of form that is sketched in Darwin's *Origin of Species*. This commonality notwithstanding, the differences between these two approaches are immense. One dimension on which they differ is historical/sociological. Throughout much of the Twentieth Century, replicator biology has enjoyed the privilege of being the sole heir to Darwinism. A continuous historical line can be sketched from Darwin's theory to current replicator biology. So completely has replicator thinking come to dominate our evolutionary thinking that the vast differences between it and Darwin's theory often go unremarked. Replicator biology is a compelling and powerful successor to Darwin's theory, but it is also a significant extension. Its methodological and metaphysical commitments go far beyond Darwin's. Only in recent years has the prospect of a substantially different extension to Darwin's theory begun to emerge.

In this essay I have attempted to articulate the core commitments of replicator theory and contrast them with what I take to be the core commitments of a thorough-going organism-centered approach to evolution. They differ in what each takes to be the canonical unit of biological organisation. As a consequence of this they differ in their respective accounts of the relation between Darwin's two key causes of evolution: 'struggle for life' and the conditions of existence. Furthermore, they differ on the most appropriate explanatory mode in which to explain the higher-order effect of adaptive evolution.

Replicator Biology takes replicators—roughly speaking 'genes'—to be the canonical units of biological organisation. Replicators take pride of place, on this approach, because they unify what are regarded as three discrete component processes of evolution: inheritance, development and population change. Inheritance, according to replicator theory is simply the copying and transmission of replicators. Development is simply the expression of phenotypic information encoded in replicators, and evolution is the change in the replicator structure of a population. Inheritance and development are essentially evolutionarily conservative processes. Evolutionary change comes about by the capacity of environments to retain and promote those variant replicators that promote the survival and reproduction of organisms. The ultimate source of the variants is random mutation. Form is changed by the capacity of mutation and replication to solve problems set by the conditions of existence. This, in turn, requires that the conditions of existence are essentially decoupled from the struggle for life. Because of its reliance on random mutation, replicator biology is committed to the ineluctable chanciness of evolution. The process that generates novel phenotypes—genetic mutation—is essentially random and unbiased. Evolution is 'chance caught on the wing'.

By contrast, organism-centered biology takes the organism to be the canonical unit of evolutionary change. By this is meant that it is the distinctive characteristics of organisms that cause the changes in population structure. The struggle for existence, on this approach, is constituted by a certain goal-directed capacity of organisms, to maintain viability by making compensatory changes in form and function. This capacity manifests itself both in the process of development and in inheritance. Phenotypic plasticity allows organisms to make changes in order to accommodate the vagaries of their conditions of existence. It also allows them to secure the resemblance of offspring to parent, despite the vicissitudes of genes and environment [41]. Because organisms are actively involved in changing and modulating the impact of their environments, the *Struggle for Life* partially constitutes the conditions of existence. To explain the process of adaptive evolution, according to organism-centered biology, one must take into account the way that organisms make changes to form and precisely function because they are conducive to survival. Evolution must be explained teleologically. Evolution is not irreducibly chancy, it is inherent in the very capacities of organisms manifested in their struggle for existence.

A century and a half after the publication of Darwin's magisterial work it remains both the source of scientific inspiration and scientific dispute. It is the founding document of two extravagantly different conceptions of the process of evolution. I do not know whether, or to what degree, these to conceptions of evolution are mutually compatible. The important point for our purposes is that each is consistent with the theory of evolution announced in *The Origin*. Darwin's theory of descent with modification continues to provide a framework for the study of the fit and diversity of organic form. If much of the intervening century and a half has been devoted to the single-minded development of one interpretation of Darwinism—Modern Synthesis Replicator Theory—I suggest that the next phase of the project inaugurated by the *Origin of Species* should be one in which two distinct, alternative, and perhaps mutually incompatible versions of Darwin's theory be weighed and compared. I suggest that the leading question should be which of these extensions of Darwinism offers the better account of the process that Darwin discovered.

References

1. Depew, D.: Adaptation as product and process: Protracting genetic Darwinism's past into developmental Darwinism's future. Studies in the History of Biology and the Biomedical Sciences (Forthcoming)
2. Walsh, D.M.: Situated adaptationism (Forthcoming)
3. Darwin, C.: On the Origin of Species. Penguin Classics, London 1859 [1968]
4. Pearce, T.: A great complication of circumstances – Darwin and the economy of nature. J. Hist. Biol. **43**(3), 493–528 (2010)
5. Paley, W.: Natural Theology. 1802. Oxford's World Classics, Oxford (2006)
6. Hull, D.: Darwin's science and victorian philosophy. In: Hodge, M.J., Radick, G. (eds.) Cambridge Companion to Darwin, pp. 168–191. Cambridge University Press, Cambridge (2003)
7. von Helmholtz, H.: The aim and progress of physical science. In: Kahl, R. (ed.) Selected Writings of Hermann von Helmholtz, 223–45. Wesleyan University Press, Middletown. [1869]1971
8. Hodge, M.J.: The structure and strategy of Darwin's long argument. Br. J. Hist. Sci. **10**, 237–246 (1977)
9. Sober, E.: The Nature of Selection. MIT Press, Cambridge (1984)
10. Salmon, W.C.: Causality and Explanation. Oxford University Press, Oxford (1998)
11. Walsh, D.M.: Chasing shadows – natural selection and adaptation. Stud. Hist. Philos. Biol. Biomed. Sci. **31**, 135–153 (2000)
12. Matthen, M., Ariew, A.: Selection and causation. Philos. Sci. **76**(2), 201–224 (2009)
13. Schrodinger, E.: What Is Life? Dover, New York (1944)
14. Sober, E.: Evolution, population thinking, and essentialism. Philos. Sci. **47**, 350–383 (1980)
15. Dawkins, R.: The Selfish Gene. Oxford University Press, Oxford (1982)
16. Lewontin, R.C.: The Tripe Helix: Genes, Organisms and Environments. Oxford University Press, Oxford (2001)
17. Fox Keller, E.: The Century of the Gene. Harvard University Press, Cambridge (2000)
18. Callebaut, W., Muller, G.B., Newman, S.A.: The organismic systems approach: EvoDevo and the streamlining of the naturalistic agenda. In: Sansom, R., Brandon, B. (eds.) Integrating Evolution and Development: From Theory to Practice, pp. 25–92. MIT Press, Cambridge (2007)
19. Gilbert, S.F., Sarkar, S.: Embracing complexity: Organicism for the 21st century. Dev. Dyn. **219**, 1–9 (2000)
20. Moczek, A.: On the origins of novelty in development and evolution. BioEssays **30**, 432–447 (2008)
21. Sterelny, K.: Novelty, plasticity and niche construction: The influence of phenotypic variation on evolution. In: Barberousse, A. (ed.) Mapping the Future of Biology. Boston Studies in the Philosophy of Science, vol. 266. Springer, Dordrecht (2009), pp. 93-110
22. Kirschner, M., Gerhart, J.: The Plausibility of Life: Resolving Darwin's Dilemma. Yale University Press, New Haven (2005)
23. Moczek, A.: Phenotypic plasticity and diversity in insects. Philos. Trans. R. Soc. B **365**, 593–603 (2010). doi:10.1098/rstb.2009.0263

24. Pfennig, D.W., Wund, M.A., Snell-Rood, E.C., Cruickshank, T., Schlichting, C.D., Moczek, A.P.: Phenotypic plasticity's impacts on diversification and speciation. Trends Ecol. Evol. **25**, 459–467 (2010)
25. West-Eberhard, M.J.: Developmental Plasticity and Phenotypic Evolution. Cambridge University Press, Cambridge (2003)
26. West-Eberhard, M.J.: Developmental plasticity and the origin of species differences. Proc. Natl Acad. Sci. US **102**, 6543–6549 (2005)
27. Godfrey-Smith, P.: Complexity and the Function of Mind in Nature. Cambridge University Press, Cambridge (1996)
28. Godfrey-Smith, P.: Organism, environment, and dialectics. In: Singh, R., Krimbas, C., Paul, D., Beatty, J. (eds.) Thinking About Evolution, Vol. 2: Historical, Philosophical, and Political Perspectives, pp. 25–266. Cambridge University Press, Cambridge (2001)
29. Walsh, D.M.: Two Neo-Darwinisms. Hist. Philos. Life Sci. **32**, 317–340 (2010)
30. Gibson, J.J.: The Ecological Approach to Visual Perception. Houghton Mifflin, Boston (1979)
31. Ingold, T.: Culture and the Perception of the Environment. Cambridge University Press, Cambridge (1986)
32. Walsh, D.M.: Teleology. In: Ruse, M. (ed.) The Oxford Handbook of Philosophy of Biology, pp. 113–137. Oxford University Press, Oxford (2007)
33. Monod, J.: Chance and Necessity. Penguin, London (1971)
34. Mayr, E.: Towards New Philosophy of Biology. Harvard University Press, Cambridge (1976)
35. Dawkins, R.: Universal Darwinism. Reprinted In: Hull, D., Ruse, M. (eds.) Oxford Readings in Philosophy of Biology. Oxford University Press, Oxford (1998)
36. Mayr, E.: Cause and effect in biology. Science **131**, 1501–1506 (1961)
37. Ariew, A.: Mayr's ultimate/proximate distinction reconsidered and reconstructed. Biol. Philos. **18**, 553–565 (2003)
38. Lewens, T.: What's wrong with typological thinking? Philos. Sci. **76**, 355–371 (2009)
39. Von Bertalanffy, L.: General Systems Theory. George Barziller, New York (1969)
40. Newman, S.A., Muller, G.B.: Genes and form. In: Neuman-Held, E., Rehman-Suter, C. (eds.) Genes in Development: Re-reading the Molecular Paradigm. Duke University Press, Durham (2007)
41. Gibson, G.: Getting robust about robustness. Curr. Biol. **12**, R347–R349 (2002)

Frequency Dependence Arguments for the Co-evolution of Genes and Culture

Graciela Kuechle and Diego Rios

Introduction

In the last few decades, a wealth of evolutionary models has been explored to account for the possible relationship between culture and genes. The seminal work of Edward Wilson [1] opened a fruitful debate that was further enlarged by new insights coming from evolutionary psychology [2], behavioral ecology [3] and dual inheritance models [4, 5]. All these models share a strong commitment to Darwinism; they disagree, nevertheless, on the details of how to carry out the Darwinian project in the social sciences. A prominent topic of disagreement concerns the causal direction of the gene-culture link. While sociobiology and evolutionary psychology provide a framework that is essentially bottom-up (from genes to culture), dual inheritance theory is ready to allow for more complex – top down – processes, where culture plays a crucial role in the fixation of genes. This paper is a contribution to this debate. We will briefly analyze one specific co-evolutionary mechanism – the Baldwin effect – that is supposed to provide a path through which culture could impinge and "direct" our genetic structures [6].

The Baldwin effect amounts to the integration into the genome of originally learnt traits. From its inception, the Baldwin effect has been regarded with skepticism. Different reasons underlie this skepticism. Some critics argue that the Baldwin effect is inevitably committed to Lamarckism [7]. These critics argue that genes are insensitive to the vagaries of acquired traits, and that they are unable to pass what has been learnt during the lifespan of an organism on to the next generation. According to this view, this commitment to Lamarckism excludes the

G. Kuechle (✉) • D. Rios
Witten Herdecke University, Witten, Germany
e-mail: diego.martin.rios@gmail.com

M. Brinkworth and F. Weinert (eds.), *Evolution 2.0*, The Frontiers Collection, DOI 10.1007/978-3-642-20496-8_15,

Baldwin effect from the set of serious evolutionary mechanisms. Other criticisms have taken a different tack, insisting on the lack of significant and well-established empirical evidence supporting the Baldwin effect. While the first objection is, we believe, mistaken, there might be much to say in defense of the second one. Although the paucity of empirical evidence is still a serious problem for any fully-fledged support of the Baldwin effect, we believe that it is not impossible to devise experimental settings to test it. We have treated the first objection and provided a possible solution, in another paper [8].

The focus of this text lies somewhere else. Despite our (moderate) optimism regarding the prospect for Baldwinian processes, the main objective of the present paper is not to address the objections raised by the critics; we will rather look more carefully at one of the arguments that are purported to support the Baldwinian project. It has been argued that Baldwinization would be particularly likely within positive frequency dependent contexts [9–11]. We claim that this thesis is incomplete. After analyzing the role of different types of frequencies in the Baldwin effect, we put forth a framework that, we believe, is able to better identify the factors governing Baldwinian processes.

In section "Baldwin's Conjecture", we introduce the Baldwin effect. In section "Generative Entrenchment and the Arithmetic of Plasticity and Fixation", we say a few words about the arithmetic of the trade-off between plasticity and fixation. In section "The Positive Frequency Argument", we present the main tenets of the positive frequency dependence account. In section "Game Theory and Strategic Interaction", we briefly explore a game-theoretic framework that, we believe, is extremely powerful to clarify some misunderstood aspects of the Baldwin effect. In section "The Critique Generalized", we go back to the positive frequency account and subject it to systematic scrutiny. The main contention of this paper is that once the game-theoretic framework is put into motion, it provides a better model to appraise the dynamics of Baldwinization than the prevalent framework that relies on the existence of positive frequencies. Last but not least, we close our investigation with some conclusions.

Baldwin's Conjecture

Baldwin's idea was an attempt to provide a rigorous Darwinian mechanism capable of accounting for the progressive genetic assimilation of learnt traits. Thanks to this mechanism, the fixation of originally acquired traits could be done in purely selectional terms. The best way to conceptualize Baldwin's thought is by means of an example. Imagine a population where an advantageous trait emerges – a new way, for instance, of climbing a tree and gathering nutritious fruits from its branches. Assume, furthermore, that this trait spreads through the population by social learning. Now, any genetic mutation that facilitates the acquisition of the new trait will be selected for, and in the long run, the trait that was originally acquired by costly trial and error will be progressively integrated into the genome.[1]

As the previous example shows, the Baldwin effect is an ingenious way of accounting for the emergence of instincts and innate structures: organisms strive to cope with the environment, learn new skills, and these fit-enhancing skills recruit favorable mutations that fix the originally acquired trait. The basic idea is that the existence of learnt traits is "guiding" – although in an indirect and unforeseen way – the whole process toward genetic fixation. Note that genetic structures are a consequence rather than a cause of the evolutionary process. The generation of mutations is still random, but their retention is not: the prior existence of fit-enhancing skills *biases* the transmission of certain mutations [9].[2]

We have described the Baldwin effect in terms of the genetic take-over of originally learnt traits. This characterization might lead to misunderstandings. It is important to keep in mind that the mechanism envisaged by Baldwin is essentially populational and selectional, not developmental. This is a striking difference with Lamarckian evolution. Within the Lamarckian picture, phenotypic innovations impinge directly on the heritable make-up of the organism, hereby passing onto later generations. In the case of the Baldwin effect, the process is a bit more complex because the path leading to genetic fixation provides not only a crucial role to learning, but also to random mutations. The first generation of adopters does not change their current genetic make-up due to having adopted the new skill. Genetic changes arise only in later generations; these genetic changes are the result of intergenerational random mutations.

Many examples of Baldwinization commonly discussed in the literature rely on an intermediate stage of social learning [9]. Note, nevertheless, that social learning is not a necessary condition for Baldwinization: asocial learning might also lead to genetic fixation. Filial imprinting has been interpreted along these lines. It has been argued that mother-following behavior was, at the beginning, acquired through trial and error; ancestral chicks wandered around to find their own mothers [12, 13]. Those organisms having discovered, through individual trial and error, that movement is a statistically reliable cue of parental presence, would enjoy a comparative advantage. Thereafter, any genetic mutation facilitating the adoption of this behavior would be selected for. In this example, there are no grounds to believe in social learning. Yet this does not preclude the possibility of Baldwinian evolution, provided that those chicks having learnt the trick flourish in the population. What happens in the next generation? There is an over-representation of chicks whose parents have learnt the new trick. If they happen to have, by sheer chance, a new mutation facilitating the acquisition of the new trait by trial and error, they will be selected for. Note, nevertheless, that it is thanks to having learnt the new trait, that the favorable mutation is retained. If the chicks had failed to adopt the adaptive

[1]The Baldwin effect is not limited to behavioral traits as opposed to physiological traits. Environmentally-induced physiological adaptations also count as an instance of Baldwinization.

[2]Note that mutations are not strictly necessary. Hidden variations may suffice for assimilation.

behavior, the favorable mutation would have been lost in the long run. This means that the intermediate stage of learning is crucial in this case. Social learning certainly contributes to Baldwinization, but asocial learning might also deliver the Baldwinian goods.

We have pointed out that social learning is not required. Yet the most compelling examples for Baldwinization do involve social learning. This needs an explanation. In fact, there are two reasons for which social learning is particularly important. The first one concerns time. Social learning strongly facilitates Baldwinian evolution because it contributes to the spread of advantageous traits. The essentially contagious nature of social learning increases the chances of adaptive behavior to remain in the population for longer time; this leaves room for selection to update the genetic base. Random genetic mutations are, in this way, allowed to track the advantageous phenotypic explorations undertaken by the organism.

The second mechanism is even more important. Social learning generates a spill-over effect. The easy transmission of an advantageous behavior through social learning parses the entire population in search for already existing individuals with the adequate genetic material. Whenever an individual with the favorable mutation is found in the population, it will have a great advantage in acquiring the new skill; in the long run these lucky organisms will turn out to have more descendants. Note that having the favorable mutation is not fit-enhancing in itself. It is also necessary to display the relevant behavior. The possibility of social learning drastically increases the chances of acquiring this behavior. Without social learning, those individuals having favorable mutations will need to rely on producing the adaptive behavior by sheer luck or by individual learning.

To sum up: although both social and asocial learning might work, the result will certainly be less robust and stable in the case of asocial learning than in the presence of social learning. Asocial learning has a very weak role to play in what concerns the time effect, and a null impact on the spill-over effect of the adopted trait. By definition, asocial behavior is not contagious; other organisms cannot copy it, and consequently the parsing of the population in search of favorable mutations is precluded. Given that the Baldwin effect provides a paramount role to the adopted trait in the recruitment of suitable favorable genetic bases, the fact that behavior cannot disseminate easily across the population – due to the lack of social learning – seriously hinders the entire process.

Generative Entrenchment and the Arithmetic of Plasticity and Fixation

An interesting way of conceptualizing the Baldwin effect is in terms of generative entrenchment. A generative entrenched structure is one that has many other things depending on it [14]. Spoken language, the advent of written and alphabetic languages and the dissemination of farming and agriculture are deeply entrenched adaptations: many further inventions and innovations rely on them [15]. We claim that the notion of generative entrenchment can fruitfully illuminate the mechanics of the Baldwin effect.

Wimsatt points out that generative entrenchment contributes to *frozen* accidental adaptations making them less prone to change and perturbation [14]. In this way arbitrary traits or phenomena can become necessary. Generative entrenched structures produce their own reinforcement through a cumulative scaffolding process on which novel traits or phenomena become progressively fossilised. A distinctive feature of generative entrenchment is that it is resilient to change. Interestingly, the stability of generative entrenchment is both at the origin of the process as well as at the end. Generative entrenchment is essentially a feed-back process that takes stable phenomena as inputs, building up further adaptations that precisely reinforce the stability of the original phenomena. This process is particularly salient in the case of the Baldwin effect. Once an adopted behavior is partially assimilated, it increases the performance of the target behavior, further stabilizing it.

For our purposes, it is important to insist on input-stability rather than on output-stability. In fact, input-stability is essential for the Baldwinian entrenchment mechanism to work: it provides a platform for further opportunistic improvement throught selection. For genes to be able to follow behavior, the inputs must be sufficiently robust and resilient to allow for genetic updating; transient behaviors simply do not leave enough time for genes to operate; they will *wash away* before genetic assimilation has had time to take place. Allowing time for genetic fixation is then a crucial factor in Baldwinian evolution, and the stability of the input-behavior is crucial to secure this requirement. Although time is then decisive, the Baldwin effect does not require that the assimilated trait remains adaptive forever. It just requires that the adopted behavior remains adaptive until genetic assimilation takes place. Once the trait has been saved into the hardware, it need not be adaptive anymore. As it will become apparent later on, this feature of Baldwinian evolution will play an important role in the development of our argument.

The process of generative entrenchment might be exposed to severe tradeoffs. Godfrey-Smith pointed out that the strategy of fixing behavior may become maladaptive when the environmental conditions affecting its fitness change [16]. The convenience of adopting a behavior that is responsive to the environment and therefore flexible, depends on multiple factors. First, the difference in payoffs of a behavior in the best and the worst state of the world. Second, the probability of each state of the world; and third, the existence of a cue capable of providing reliable information about the true state of nature. The main conclusion is that if the organism lacks a way of tracking the environment, it is always better to fix behavior.

There is a strong intuition as to the importance of environmental change and stability as requirements for flexible behavior and learning. Without change, there is no need for adopting a flexible behavior. Too much change, on the other hand, will likely render any attempt to find a regularity obsolete. Stephens [17] distinguishes environmental change within generations and environmental change between generations, and concludes that learning is favored when within-generation predictability is high and between-generation predictability is low. Under these conditions, early experimentation allows the organism to adapt to the conditions characterizing its later life. Summarizing, unstable environments act not only against genetic fixation, but also against any attempt to learn and acquire knowledge with the aim of exploiting regularities present in recurring events [13, 16]. Since the stability of the environment does not crucially bear on the arguments about the Baldwin effect that we are concerned with in this paper, we will assume that the environment is stable in the sense just described. In like manner, we will assume that the mentioned advantages of the behavior under consideration outweigh their costs, since this is a precondition for the Baldwin effect.

We have now introduced the basics of Baldwian evolution and made a few comments on the trade-off between plasticity and fixation. In the next section, we will briefly present a prominent framework aimed at identifying the conditions that would foster Baldwinization. We will later provide an alternative model that we believe better captures the conditions governing the Baldwin effect. This framework will allow us to assess the validity of the positive frequency account.

The Positive Frequency Argument

There are different ways of arguing for the Baldwin effect. One possible strategy consists of showing its feasibility. Hinton and Nowlan's simulation is an example of this kind of defence of the Baldwin effect [18]. Another possible justificatory strategy consists of identifying a range of phenomena that could be explained by the Baldwin effect in more simple, intuitive or compelling ways than natural selection simpliciter. David Papineau's complexity hypothesis is an example of this argumentative strategy [19]. In this paper we will not tackle either one of these two strategies. We will rather focus on a different attempt to justify the Baldwin effect – to wit, the positive frequency account advocated by Deacon and Godfrey-Smith [9, 10].

The objective of the positive frequency account is to identify contexts that could favor the emergence of Baldwinian evolution. Positive frequency dependence occurs whenever the advantage of having a trait increases with the proportion of individuals in the population who have a convergent trait [9, 10, 20, 21]. Consider for instance the case of language. The more people speak the language, the easier it is to find exemplars to learn the language from, and the more rewarding the acquisition of that language. There is a networking process in place with mutually reinforcing interactions. A language with few speakers is both difficult to learn and less rewarding to acquire.

The positive frequency argument is a niche-construction argument. Niche-construction refers to the diverse ways through which organisms change their own environment, thus changing the selective forces acting upon them [22]. The construction of nests, webs and burrows are rather obvious forms of modifying the environment, generating new contexts for selection to operate. "Environment" usually denotes the set of exogenous elements affecting the fitness of an organism, including not only the physical habitat but also the behavior of unrelated species. However, there is a more inclusive notion of "environment" that includes the behavior of all other conspecifics. This notion is highly relevant for individual organisms since behaviors are generally interdependent. When the fitness of an organism is frequency dependent, its environment will endogenously change with the behavior of the whole population. The positive frequency account exploits this idea to the full. The advantage of having a certain trait depends on the convergent behavior of the other organisms in the population. Within the positive frequency account, each new adopter modifies the environment of the other organisms, increasing the advantage of adopting the new behavior.

There is much to commend on the idea of positive frequency dependence. It points to an important factor that could certainly be involved in the Baldwinization process, but it cannot be the whole story. Let us explain this point. Note that positive frequency dependence is a form of increasing returns because the payoff to adoption increases with the proportion of adopters. Although several phenomena display this feature, there are many interactions characterized by constant and even decreasing returns to adoption. In the case of constant returns, the advantage of adopting a phenotype does not depend on the proportion of adopters and, in the case of decreasing returns, the advantage even decreases with the frequency of adopters. The trick to opening coconuts for instance, provides an example of this last situation: the higher the proportion of individuals who have adopted the trick, the more difficult it will be to find coconuts to open. However, adopting the trick is adaptive even if it improves the fitness of the adopters at a decreasing rate. This example shows that the relevant criterion for adoption is not determined by the type of frequency dependence, but by the best-response nature of this behavior, given what other individuals are doing.

We claim that what matters is not how the fitness of a phenotype varies with its frequency, but whether adopting a phenotype is an above-average response at the population level, given the current proportion of phenotypes in the population. It is of no direct significance whether these interactions involve increasing, decreasing or constant returns to adoption. Notice the exact nature of our argument. We do not dispute the truth of the positive frequency account. We merely claim that it fails to pinpoint with enough accuracy the factors at play in the Baldwin effect. As will be argued for in the remaining sections of this paper, this larger class might be fruitfully described in game-theoretic terms.

Game Theory and Strategic Interaction

The main advantage of the effect envisaged by Baldwin is that it saves the organism the hassle of acquiring a behavior through trial and error. The potential drawback is that, by placing a behavior under genetic control, the organism loses phenotypic flexibility which may be necessary when the environment changes.[3] Therefore we argue that any behavior with potential for Baldwinization must remain adaptive at least until genetic assimilation is accomplished. In other words, it is not enough that a certain phenotype be adaptive under the conditions currently faced by the organism. Baldwinization requires that the phenotype be able to resist the invasion of novel behaviors that may jeopardize its persistence until genetic fixation is reached. In this light, we argue that evolutionary stability is a prerequisite for Baldwinization.

Loosely speaking, a phenotype is evolutionary stable if it is fit enough to outcompete alternative phenotypes and resist their invasions when adopted by virtually every individual in a population. Evolutionary stable strategies perform better than or equal to any other mutant strategy against themselves, and if there is a mutant that performs equally well as the ESS against the ESS, then the ESS must perform strictly better than this mutant against the mutant. In this way evolutionary stable phenotypes are able to resist the invasion of every possible mutant phenotype, at least under the assumption of small and isolated mutations and large populations. Evolutionary stability is a refinement of Nash equilibrium that is relevant under certain forms of selection and replication of behaviors. We focus on this concept because of its central role in the literature but we would like to emphasize that our argument relies on the general concepts of equilibrium and dynamic stability and could be restated in terms of other stability concepts. We have dealt with this issue in more detail elsewhere [8]. To simplify matters, and given the fact that its properties are compatible with different kinds of learning and imitative behavior, we will also assume a particular type of dynamics – namely, the continuous-time version of the replicator dynamics.

For concreteness, consider a population where some individuals discover a new trick to open co-conuts.[4] Since this skill is highly advantageous, mutations facilitating the acquisition of this skill can be expected to be retained.[5] Figure 1 depicts the relative fitness of each individual for each profile of behaviors. The particular payoffs are irrelevant (we explicit them to make the example more vivid) as long as two conditions are met: first, the fitness of adopting the trick is higher when the other individual does not adopt it ($3 > 2$) – an effect that could, for instance, be caused by a fixed supply of coconuts – and second, the fitness of not

Fig. 1 Interaction with a dominant strategy

	adopt the trick	do not adopt the trick
adopt the trick	(2, 2)	(3, 1)
do not adopt the trick	(1, 3)	(1, 1)

[3]Hinton and Nowlan [18] develop a computational model that explores the interplay between the benefits and costs of learning and the impact of these variables upon adaptive evolution.

[4]This example is analyzed in Papineau [19].

[5]For simplicity, we deal with pairwise interactions instead of the more realistic set-up in which phenotypes play the field. But our argument about evolutionary stability holds in any case.

adopting the trick does not depend on the behavior of the other individuals (no adoption always yields a payoff of 1). Notice that in this game, the fitness of adopting the trick is negative frequency dependent: the higher the number of adopters, the lower the relative fitness of the adopters.[6] Despite this fact, adopting the trick dominates the alternative phenotype, and thus fares better, irrespective of its frequency in the population.

As usual in this type of games, the set of profiles that constitute a Nash equilibrium are potential candidates for evolutionary stability. These are profiles such that no player can do better by changing his or her behavior if the other conforms to it. The profile (adopt the trick, adopt the trick) is the only Nash equilibrium of this game. Moreover, since no feasible set of payoffs can make an individual better off without making the other worse off, the Nash equilibrium is also Pareto optimal. Given the definition offered in the previous paragraph it is straightforward to check that "adopt the trick" is evolutionary stable and a global attractor of any adaptive dynamics.[7] For this reason, we can confidently expect the population to adopt the trick in the long run.

By definition, Baldwinization involves phenotypes that are fit or adaptive. Although evolutionary stability also entails the requirement of adaptiveness, it imposes additional conditions. Evolutionary stable strategies are not only adaptive, but they are also able to resist invasions from other phenotypes that arise from small and sporadic mutations. It follows from our argument that not every behavior yielding above average fitness could qualify for Baldwinization so that adaptiveness as such is generally insufficient for Baldwinization. The rationale for this argument is that the Baldwin effect takes time. A phenotype that is undergoing genetic fixation but fails to be evolutionary stable, may turn maladaptive along the way. Natural selection myopically favors adaptive behaviors, i.e. behaviors that have currently performed better than average and these need not be evolutionary stable. Evolutionary stable behaviors have the potential to attract the dynamics of the system in the long run and persist. By this token, if a process of Baldwinization targets a currently adaptive but non-evolutionary stable behavior, it may not be able to finalize the process.

The requirement of evolutionary stability entails some constraints to the Baldwin effect. First, evolutionary stable strategies constitute a subset of Nash equilibria that may fail to exist. In the absence of stable equilibria, a dynamic system may cycle or fail to asymptotically converge to any resting point. Second, long-run dynamics may also converge to a Nash equilibrium which is not evolutionary stable. This last case poses a problem for the argument that requires evolutionary stability, since in this situation the system converges to a strategy

[6]The payoffs to adoption and non-adoption assume that the new technology to open coconuts allows adopters to eat more than non-adopters per unit of time and that this larger intake enhances their reproductive survival.

[7]Adaptive dynamics are dynamics by which strategies whose fitness is higher than the average fitness of the population, grow.

that is not evolutionary stable. An instantiation of such a case occurs in the presence of neutrally stable strategies that, by definition, are not evolutionary stable. These strategies cannot be invaded and displaced because no mutant earns a higher payoff against them. Yet they are not fit enough to eliminate other mutants. As long as a mutant is equally fit, it may be able to coexist with them. In this light, and from a more general perspective, taking into account the plethora of factors shaping evolutionary dynamics, our argument boils down to stress the importance of stability criteria in Baldwinization processes.

In a nutshell, we have shown that the positive frequency account fails to explain canonical examples of the Baldwin effect involving decreasing returns to adoption, such as the trait of adopting the trick to opening coconuts. Acquiring a new trick to open coconuts is an advantageous behavior regardless of the behaviors of other individuals and despite the fact that its relative fitness decreases with the frequency of adopters. The positive frequency dependence argument cannot at all capture the significance of this latter example as a potential case of Baldwinization. Nevertheless, if we think in terms of best-response, as we are suggesting here, we might plausibly argue that the adoption of the trick is a good candidate for evolution through Baldwinization.

The Critique Generalized

In the previous section we claimed that the existence of positive frequency dependent behaviors is not a necessary condition for Baldwinization. In order to support this claim, we provided a counterexample: the game staged in Fig. 1 has a phenotype that meets the required stability conditions for Baldwinization, although it does not exhibit positive frequency dependent strategies. In this section, we further explore the relationship between positive frequency dependence and evolutionary stability to fully assess its impact upon Baldwinization.

To this intend, we address the question of what to expect in games characterized by more than one positive frequency dependent behavior. In this category we find coordination games. These games are characterized by the fact that every behavior is advantageous as long as the other player also adopts it.[8] Such a case is illustrated in Fig. 2 and represents the situation in which two individuals wish to communicate with one another while lacking a commonly agreed language. Players' payoffs

Fig. 2 Too many positive frequency dependent behaviors

	L_1	L_2
L_1	(1, 1)	(0, 0)
L_2	(0, 0)	(1, 1)

[8]Signaling games, pioneered by Lewis [23] to analyze conventions and the emergence of meaning, and the adoption of a technology standard are also coordination games. See Skyrms [24, 25] for an extensive treatment.

depend only on whether communication takes place and not on the peculiarities of the used language. This game has two evolutionary stable equilibria corresponding to the scenarios in which both individuals coordinate by choosing an identical language. Both equilibria have the same basins of attraction so that there is a 50% chance that the dynamics will evolve towards either of them. Which equilibrium will be selected depends on the initial conditions and on the history of the dynamics. Even if the system spends long periods of time in the same equilibrium, providing enough time to eventually complete Baldwinization, it is clear that positive frequency dependence alone is insufficient to determine which language will be adopted.

In a nutshell, positive frequency dependence is not rich enough to capture the subtleties of evolutionary dynamics. Were the payoffs to (L_1, L_1) be changed to (10, 10), stability accounts and positive frequency dependence would provide different predictions. While we would say that L_1 is a putative target for Baldwinization (having a considerably larger basin of attraction), the positive frequency account would not be able to make this prediction without being committed to an extraneous metric distinguishing degrees of positive frequency dependence.

Huttegger [26] shows that in games with more than two signals and two acts, where states are equiprobable, the replicator dynamics may become locked in suboptimal equilibria (babbling equilibria). Studying the trade-off between plasticity and canalization, Zollman and Smead [11] pit strategies that respond to environmental cues with a reinforcement learning mechanism against strategies that produce a fixed behavior regardless of the environment. In their simulation the emergence of plastic types paves the road for further fixation. However, these plastic phenotypes are later on displaced by types who fix their behavior. Zollman and Smead state that only a few games display the coexistence of plastic and non-plastic behavior.

Another interesting case of positive frequency dependence is given by the Stag Hunt game which is a prototype of the social contract [27]. The game is depicted in Fig. 3. Individuals must decide between hunting stag, which exhibits increasing returns to adoption, and hunting hare, which entails no strategic risk. The choice of "stag" is advantageous only if the other individual joins the hunt, whereas the choice of "hare" implies no need of further help. From an evolutionary game theoretic perspective, the prospects of getting individuals to choose the "stag" equilibrium are dire. Although both phenotypes are evolutionary stable (their basins of attraction depend on the particular payoff structure), if the dynamics allow large and non sporadic mutations, "hare" stochastically dominates "stag". Nonetheless, if individuals manage to find others who also hunt stag, a situation that naturally arises when interactions are spatially limited, fixation of "stag" can then spread and

	S	H
S	(4, 4)	(0, 3)
H	(3, 0)	(3, 3)

Fig. 3 Positive frequency dependent behavior is risk dominated by another behavior

prosper in the population [28]. Given that "stag" is the only positive frequency dependent behavior, the positive frequency account is committed to fix it. We would, nevertheless, rather remain agnostic about the final outcome.

Our argument for the Baldwin effect must not be overstated. Beyond the case of dominant strategies and socially optimal equilibria, Baldwin effects become more dependent upon the nature of the shocks that perturbate the system.[9] In games with multiple evolutionary stable strategies, if the system is subject to small and sporadic perturbations, evolutionary stability may still enable Baldwinization. If, on the other hand, perturbations are large and recurrent, the dynamics may eventually tip the system from one equilibrium toward the other. The likelihood of the Baldwin effect would in these cases depend upon the amount of time that the system spends in each equilibrium. Lastly, it is worth noticing that the games analyzed here drastically abstract from the complexity that characterizes the structure of biological organisms that may impose considerable constraints upon the given evolutionary dynamics.

Conclusions

It is time now to briefly summarize the conclusions of this paper. We discussed a prominent justification of the Baldwin effect – to wit, the positive frequency dependence account. According to its advocates, the positive frequency account is able to pinpoint the conditions that provide fertile grounds for the Baldwin effect. We argued that this claim does not capture the factors involved in Baldwinization; positive frequency phenotypes are neither sufficient nor necessary for the Baldwin effect. We provided a game-theoretic analysis based upon the notions of equilibrium and stability – to assess the possibility of Baldwinization. To this intent, we contended that social interactions, whose evolutionary stable strategies involve dominant phenotypes, provide paradigmatic grounds for Baldwinization, especially when the equilibrium is socially optimal. We showed that these factors apply even in the presence of negative frequency dependence.

References

1. Wilson, E.: Sociobiology: The New Synthesis. Harvard University Press, Cambridge (1975)
2. Barkow, J.H., Cosmides, L., Tooby, J. (eds.): The Adapted Mind. Oxford University Press, Oxford (1992)
3. Krebs, J., Davies, N.: An Introduction to Behavioral Ecology. Blackwell, Oxford (1982)
4. Boyd, R., Richerson, P.: Culture and the Evolutionary Process. The University of Chicago Press, Chicago (1985)

[9] As stated above, the rules guiding the interaction of the individuals, most prominently the type of spatial interaction, will also affect the evolutionary dynamics of those games.

5. Durham, W.: Co-evolution. Stanford University Press, Stanford (1991)
6. Baldwin, J.: A new factor in evolution. Am. Natl. **30**, 441–451, (1896)
7. Watkins, J.: A note on the Baldwin effect. Br. J. Philos. Sci. **50**, 417–423 (1999)
8. Kuechle, G., Rios, D.: A game-theoretic analysis of the Baldwin effect. Erkenntnis (forthcoming)
9. Deacon, T.: The Symbolic Species. Norton, New York (2007)
10. Godfrey-Smith, P.: Between Baldwin skepticism and Baldwin boosterism. In: Weber, B., Depew, D. (eds.) Evolution and Learning: The Baldwin Effect Reconsidered. MIT Press, Cambridge (2003)
11. Zollman, K., Smead, R.: Plasticity and language: an example of the Baldwin effect? Phil. Stud. 147:7–2 (2010)
12. Ewer, R.: Imprinting in animal behaviour. Nature **177**, 227–228 (1956)
13. Avital, E., Jablonka, E.: Animal Traditions: Behavioral Inheritance in Evolution. Cambridge University Press, Cambridge (2000)
14. Wimsatt, W.: Re-engineering Philosophy for Limited Beings. Harvard University Press, Cambridge (2007)
15. Diamond, J.: Guns, Germs and Steel. Norton & Company, New York (1997)
16. Godfrey-Smith, P.: Complexity and the Function of Mind in Nature. Cambridge University Press, Cambridge (1996)
17. Stephens, D.: Change, regularity and value in the evolution of animal learning. Behav. Ecol. **2**, 77–89 (1991)
18. Hinton, G., Nowlan, S.: How learning can guide evolution. Compl. Syst. **1**, 495–502 (1987)
19. Papineau, D.: Social learning and the Baldwin effect. In: Zilhao, A. (ed.) Evolution, Rationality and Cognition: A Cognitive science for the Twenty-first Century. Routledge, Abingdon, Oxon (2005)
20. Suzuki, R., Arita, T.: How learning can guide the evolution of communication. *Proceedings of Artificial Life XI*, pp. 608–615 (2008)
21. Suzuki, R., Arita, T., Watanabe, Y.: Language, evolution and the Baldwin effect. Artif. Life Robot. **XII**(1), 65–69 (2008)
22. Odling-Smee, J., et al.: Niche Construction: The Neglected Process in Evolution. Cambridge University Press, Cambridge (2003)
23. Lewis, D.: Convention. Harvard University Press, Cambridge, MA (1969)
24. Skyms, B.: The Evolution of the Social Contract. Cambridge University Press, Cambridge (1966)
25. Skyms, B.: Signals, evolution and the explanatory power of transient Information. Philos. Sci. **69**, 407–428 (2002)
26. Huttegger, S.: On robustness in signalling games. Philos. Sci. **74**, 839–847 (2007)
27. Skyrms, B.: The Stag Hunt and the Evolution of Social Structure. Cambridge University Press, Cambridge (2003)
28. Alexander, J.M., Skyrms, B.: Bargaining with neighbors: is justice contagious? J. Philos. **96**, 588–598 (1999)

Taking Biology Seriously: Neo-Darwinism and Its Many Challenges

Davide Vecchi

Introduction

Celebrating the tremendous influence that Darwin's ideas have on our culture is surely legitimate. It is difficult to underestimate the enormous influence that the idea of evolution by natural selection has exerted on many branches of philosophy and science. The idea is so simple and so powerful that, as it has been repeatedly noticed, on the one hand everyone thinks to understand it correctly, and, on the other, as a consequence of this confidence misinterpretations and misapplications abound, especially in the form of attempts to explain in some putatively novel fashion a particular phenomenon of change. Gould [1] and Fracchia and Lewontin [2] tried, perhaps with too much scepticism, to refrain scholars from exploiting the strength of Darwin's idea. They warned that the idea works for biological evolution, but not for other evolutionary phenomena. Their anti-conformist warning was meant to avert a dangerous trend to transform Darwinism into a cottage industry. The rise to public notoriety of evolutionary psychology and evolutionary psychiatry are examples of such over-exploitation of Darwin's legacy.

I personally do not share this kind of scepticism. For instance, evolutionary approaches to culture are generating some interesting ideas, even though perhaps not novel explanations (one has only to check the recent literature in evolutionary archaeology and linguistics). Furthermore, as a philosopher I might be excused to be satisfied with the heuristic power of a good analogy. After all, evolutionary epistemology, as Popper characterised it, was not much more than an analogy, and philosophers are still interested in Popper – in the philosophy of biology mostly as a critical target. Only time will say whether the sceptics are correct.

As I see it, the problem is not whether Darwin's idea is applicable to culture (I take it for granted and established), but it is rather to assess whether and to what extent these

D. Vecchi (✉)
Philosophy Department, Universidad de Santiago de Chile, Santiago, Chile
e-mail: davide.s.vecchi@gmail.com

M. Brinkworth and F. Weinert (eds.), *Evolution 2.0*, The Frontiers Collection,
DOI 10.1007/978-3-642-20496-8_16,

evolutionary approaches to culture are Darwinian: to use phylogenetic methods in order to analyse patterns of artefact distribution or linguistic usage does entail believing that selection has solely shaped such patterns? The answer is obviously no, because other factors can cause evolution. The interesting question is therefore to understand whether the idea of evolution by natural selection suffices to explain cultural adaptation and diversity. Again, this is an empirical issue and only time will give an answer to this question. However, I suspect that there are reasons for believing that what is needed is not a mere *Darwinization*, but rather a *biologization*, of culture. In this contribution I will try to clarify what I mean by focusing on new trends in biology. History shows that, in biology but not only, there has been a tendency to accept too uncritically an appealing but rather simplified vulgate of Darwinism. But things are starting to change, at least in biology, where there is growing awareness that neo-Darwinism is losing its tight grip on the reins of the community of practitioners. In this paper I will highlight some of the respects in which neo-Darwinism is increasingly seen, within the life sciences, as providing an oversimplified picture of how evolution really works.

Neo-Darwinism and Darwin's Church

Historically, the term neo-Darwinism was coined by George Romanes to refer to Weissman's ultra-selectionist ideas about evolution. Without trying to delve into surely interesting, but in this context irrelevant, exegetical issues, we can say that Darwin's Darwinism was more pluralistic than Weissman's. For instance, after Fleming Jenkin proposed the so-called "swamping" argument, Darwin came to stress Lamarckian aspects like use and disuse, direct induction and habit as determinants of heritable variation, each of them much more likely than blind variation to produce simultaneous generation of similar variants and therefore lead to significant adaptive evolution. A pluralistic Darwinism like Darwin's is sought by many biologists these days, one that emphasises the relevance of evolutionary processes beyond (or besides) selection.

Today by neo-Darwinism we do not refer to Weissman's views anymore, but rather to the Modern Synthesis' vulgata of Darwinism. This interpretation, many biologists feel, should be abandoned in favour of an account that acknowledges the profound implications of recent developments in biology. But opinions, quite naturally, differ. For instance, this is what Mayr [3] claimed:

> By the end of the 1940s the work of the evolutionists was considered to be largely completed, as indicated by the robustness of the Evolutionary Synthesis. But in the ensuing decades, all sorts of things happened that might have had a major impact on the Darwinian paradigm. First came Avery's demonstration that nucleic acids and not proteins are the genetic material. Then in 1953, the discovery of the double helix by Watson and Crick increased the analytical capacity of the geneticists by at least an order of magnitude. Unexpectedly, however, none of these molecular findings necessitated a revision of the Darwinian paradigm – nor did the even more drastic genomic revolution that has permitted the analysis of genes down to the last base pair. It would seem justified to assert that, so far, no revision of the Darwinian paradigm has become necessary as a consequence of the spectacular discoveries of molecular biology.

Many biologists would not subscribe to this view. The reasons are varied. The last 30 years of biological research have shown that, for instance, variation generation can be targeted in a variety of ways, and that it can be abrupt, systemic and even partially saltational. Molecular data concerning genome organization, genomic change, cellular and developmental processes are at odds with the anachronistic view of evolution still largely in vogue. Neo-Darwinism is challenged in a variety of respects. Overall, these challenges aim to show that evolution involves a panoply of processes, and that the role of selection, though crucial, has been overstressed.

However, neo-Darwinism remains very influential. According to one of its latest official formulations [4], evolution results from the slow replacement of one gene by another that confers a tiny reproductive advantage, where the major mode of variation generation is DNA-based mutation. The focus is entirely on selection and DNA sequences, while change is random and gradual. Coyne's position is remarkably conservative, analogous to Dawkins' [5] as advertised in the piercing *The Blind Watchmaker* back in 1986: evolution is nothing more than the process of accumulating small change via cumulative selection. But if Dawkins could have been excused somehow for not facing to the emerging facts and discoveries cropping up thanks to molecular studies, Coyne's perseverance to defend such an orthodox outlook remains puzzling to many practitioners. It seems as if here we are facing a fundamentalism that is more akin to religious fervor: neo-Darwinism as Darwin's Church.

Many believe that there should be no place for this additional form of fundamentalism. It is anachronistic because the life sciences, propelled by the advances spurred by the molecular revolution, are moving in a different direction. In this paper I will argue that if the knowledge emerging from genomics, developmental and cell biology is not systematised and incorporated in a new pluralistic theory of evolution, then we will both continue to define Darwinism simplistically (paying a limited tribute to Darwin), and, additionally, pay a disservice to the human sciences (as they will be unable to exploit the full arsenal of significant evolutionary analogies already available).

Mapping the Future of Biology

Of course, we should be critical of any approach that calls for a radical re-interpretation or even abandonment of the neo-Darwinian perspective. What many practitioners seem to be opposing is rather the result of a long-going and multifaceted process of "hardening" of the Modern Synthesis' interpretation of Darwinism, as already chronicled in some respects by Gould [6]. However, it should be noted that, historically, some neo-Darwinians have been more open-minded than others, and that Dawkins' version of neo-Darwinism remains fringe despite its popular success [7]. Furthermore, it is clear enough that we are not on the verge of a Kuhnian revolution in biology, the essential reason being that Darwin got it fundamentally right: life on earth is diverse but interrelated via common ancestry, and it evolves by natural selection. No sensible biologist would deny that selection happens and that it is real. But many people would add that something more happens, as I will try to show in the rest of the paper. Biology is in need of an extension, and the reasons to celebrate Darwin's genius remain intact.

Needless to say, there is disagreement concerning the type of revisions and extensions required by the neo-Darwinian paradigm. Gaps to be filled seem to be ubiquitous. In this paper I will focus on three areas of research that promise to contribute extensively to the emergence of a new biology: evolutionary developmental biology, microbiology and virology. Research in these areas is showing at least three putative limits of the neo-Darwinian perspective. First, contrary to what neo-Darwinians advocate, new variations are not solely produced by accidental changes in DNA sequences; on the contrary variations can be targeted in a variety of ways. Secondly, contrary to what neo-Darwinians usually believe, hereditary variation can be affected by the developmental history of the organism. Thirdly, contrary to what neo-Darwinians advocate, evolution is not only based on vertical descent, while the pattern is not always tree-like, unless we focus on a very biased selection of organisms.

The challenges to neo-Darwinism brought by the areas of biological research on which I will focus are targeted to particular aspects of the idea that numerous, successive, slight and random modifications of form will lead to evolution, adaptive complexity and biological diversity. More generally, we could say that what critics of neo-Darwinism deny is that the numerous, successive, slight and random modifications hypothesised by Darwin suffice to explain evolutionary patterns. How revolutionary this denial is remains an open question.

Evolutionary Developmental Biology: A Postmodern Synthesis?

Evo-devo is a buzzword in philosophy of biology. Evolutionary developmental biology is a continuously growing field of research with many dedicated journals and a plethora of varied publications [8]. The field has generated a lot of philosophical attention because of some easily identifiable reasons. First and foremost, because it aims to show that development contributes to evolution, potentially filling the gap between evolutionary and developmental studies opened up by the Modern Synthesis. Even though embryology has a rather obvious natural place in biology, historically developmental biology was substantially left out of the Modern Synthesis. To cut and simplify a long story short, the discovery of HOX genes in the early 1980s has revolutionised the field because it became immediately apparent how developmental processes could have evolutionary significance. Once Gould [9, p. 189] – one of the most influential critics of the Modern Synthesis as well as one of the most ardent proponent of a new synthesis between developmental biology and population genetics – wrote that evolution could work via the effects of "... small mutations with large impact upon adult phenotypes because they work upon early stages of ontogeny and can lead to cascading effects throughout embryology." This passage looks almost prescient in the light of what happened since the discovery of HOX genes in 1983. Nobody could have predicted the amount of DNA sequence conservation at the molecular level that was soon to be discovered, nor the control range of the few HOX genes on developmental processes in different and distantly related species of Metazoa. The discovery that a few genes control basic and crucial embryological processes in all animals offers almost immediately an evolutionary recipe in the form of heterochronic changes (i.e. changes in the timing of developmental events), exactly as Gould predicted.

A very significant recent debate between developmental geneticists and evolutionary biologists concerns the locus of genetic evolution: while the latter generally tend to consider evolution at base sequence level as primary (Hoekstra and Coyne [10]), the former tend to focus on changes to gene regulatory machinery (Carroll [11]). The safe answer seems to be that both modes are common and important, which means that evo-devo is partly correct in its focus. Developmental genetics is surely the area of research in evo-devo that has delivered at first, even though genes are not the sole focus in evo-devo (I will return on this point below). Developmental genetics has produced a series of interesting discoveries that have even somehow resurrected notions of systemic mutation until very recently ridiculed. Evo-devo has legitimised the appeal to saltational hypothesis and modes of evolution. For instance, Minelli et al. [12] recently discovered a centipede species (*Scolopendra duplicata*) that has twice the number of segments of its closest relative. The authors argue that morphological and phylogenetic evidence support the hypothesis that such a drastic change in segmentation is due to a very simple developmental mechanism of duplication. In this case there are no intermediate forms between this species and its closest relative, there is no gradual change but instead a jump in phylogeny, or rather a developmentally mediated "macromutation". The ensuing pattern of phenotypic evolution is somehow saltational. To put it provocatively, *Scolopendra duplicata* could be considered a "hopeful monster" in this view, to use Mayr's famous expression of abuse for Goldschmidt's ideas.

Historically, philosophical interest in evo-devo was focused on the issue of the nature of constraints. The neo-Darwinian picture of the evolutionary process was based on the idea that variation is isotropic, that is, that it is equiprobable in all possible directions around the existent phenotype [6]. This idea found an extreme formulation in Dawkins [13], who provocatively seemed to argue that selection is so powerful that, given sufficient time, any variant will appear in the gene pool. Sober [14] countered this point by pointing out that it is difficult to imagine future zebras growing machine guns to fight predators. Evo-devo has made it abundantly clear that selection is not so powerful, and more interestingly that variation is not isotropic, but rather developmentally constrained in a multifaceted manner [15]. The "constraint versus selection" debate [16] led to the realisation that developmental dynamics play a fundamental role in the generation of variation. However, even today, despite this agreement, the philosophical issue is framed in terms of the creative role played by developmental and generative processes on the one hand and selection on the other.

The central issue of contention concerns the creative role of selection in shaping evolutionary patterns. Developmentalists generally argue that selection is a pruning process, essentially getting rid of the less fit phenotypes. Selection in this view can be said to be creative only in a limited sense, namely in enabling the conditions for the emergence of innovation [17]. Selectionists sometimes reply by reframing the issue as if it were semantical. As Ruse [18] rebuts, using a possibly misleading analogy: "When Michelangelo sculpted David, was he pruning or creating?" To put it in different words, one could articulate the two positions as having two different focuses: while selectionists are concerned about the processes leading to the survival of the fittest, developmentalists focus on the processes leading to the *arrival of the fitter*. Both positions are important, even though, again, evo-devo's stance is vindicated: if variation is not isotropic then the mechanisms generating variation have evolutionary relevance.

More substantial biological issues are at stake in this debate. One concerns the proper temporal role of selection. Developmental biologists tend to see selection as mainly stabilising, as primarily acting by weeding out less fit phenotypes after they have already emerged, as ancillary and subsidiary to the genuinely creative aspect of the evolutionary process, which is creating new forms, that is, new morphologies, physiologies and behaviours. Creativity belongs to a different level that is only consequently touched by natural selection (i.e. through the fine-tuning of developmental systems). In fact, it is up to developmental mechanisms to generate evolutionary innovation. And here evo-devo approaches widely diverge, as different theorists tend to emphasise the creativity of different aspects of ontogeny: generalising perhaps too much, one could say that while most tend to stress embryological processes, other highlight the fundamental contribution of post-embryonic dynamics (including behaviour, processes of physiological adaptation and developmental reorganisation). A further rough generalisation that could be drawn concerns the origin of developmental novelty: while some developmentalists stress the internal structural properties of the developmental system (e.g. the organism), other tend to emphasise environmental influences. Of course, these rough generalisations are not aimed at producing a faithful categorisation of the variety of approaches in evo-devo. I use them in order to clarify to the reader the large variety of evo-devo's contributions. As a matter of fact, ecumenical approaches stressing the influence of both internal and external factors, as well as highlighting the evolutionary role of embryological together with processes of physiological adaptation, are the norm rather than the exception. The structuralist and environmental tradition are both well represented in evolutionary developmental biology.

Furthermore, evo-devo is naturally receptive to lines of research that focus on aspects of ontogeny that elude embryological development such as niche construction and Baldwin effect evolutionary scenarios. The traditional focus of these research areas has been behaviour. For instance, Popper [19] used both approaches in order to propose his revisions of neo-Darwinism: behaviour for Popper was the "spearhead" of evolution, the starting point of the evolutionary process (Popper can be seen as a developmentalist in this limited sense, as he puts ontogeny at the centre of the evolutionary process); new behaviours offer new opportunities and, if successful, produce changes in the environment by creating new niches (organisms, with their preferences, construct new niches); only subsequently will selection work by favouring organisms whose genomes facilitate the reconstruction of the new advantageous behaviour. Popper argued that it is this process of phenotypic evolution followed by genetic assimilation that produces significant adaptive complexity rather than neo-Darwinian mutation followed by selection.

The evolution of lactose tolerance seems to be an instance of this process. When human populations started to pass from a hunter-gatherer to an agricultural way of life, eventually raising and breeding livestock, the opportunity of consuming milk and milk products emerged. Consuming milk can be equated to a novel phenotype, somehow chosen by some of our ancestors for obvious nutritional reasons. A new niche (what we could call the milk "industry") was carved by human novel behaviour. Processes of niche construction and phenotypic evolution were followed by the genetic assimilation process: given variation within human populations concerning the capacity to metabolise lactose, selection started to favour humans with increased lactase (i.e. the enzyme necessary to digest milk) activity. Today lactose tolerance is almost endemic in northern Europe (where milk consumption was and remains high), while lactose intolerance is almost endemic in parts of Asia (where milk consumption was and is low).

This phenotypic pattern of evolution, clumsily proposed by Popper as a panacea to counter the epistemic limits of neo-Darwinism, has recently risen to new prominence, in a much more articulated version, thanks to the contributions of the biologists West-Eberhard [20] and Kirschner and Gerhart [21]. These approaches, by emphasising the role of phenotypic evolution, challenge neo-Darwinism in additional ways. Let me illustrate this point more clearly (in doing so it will also become clear a second substantial biological issue at the heart of the debate regarding the creativity of selection). Neo-Darwinians generally assume that variation is genetic in origin. Mutation is assumed to be the process providing the raw material upon which selection acts. Some evo-devo approaches challenge this orthodox view by contending that usually evolution is neither gene-based nor mutation-based, at least initially. The alternative lies in focusing on a biological phenomenon that has been neglected by mainstream evolutionary thinking: phenotypic plasticity, that is, the ability of the organism to produce a phenotype in response to its environment. If organisms are plastic then developmental processes become important if the phenotypes they produce become heritable. And this is considered to be a problem because developmental changes to the phenotype, being somatic, cannot be easily transmitted to future generations. Is this truly the case?

Before answering this question let me first note that evo-devo denies that variation is "raw", meaning of small effect and without direction. With raw variation selection is very powerful, but with rich and structured variation, as it results, for instance, from processes of developmental reorganisation and physiological adaptation, selection becomes less creative. The example of the two-legged goat is frequently cited by developmentalists to explicate the notion of developmental reorganisation. Without delving too much into the details of the case, a couple of things can be said in this context. First, the example does not show that two-legged goats are fitter than “normal” goats, and neither that goat bipedalism is a novelty from an evolutionary point of view. The significance of the example is subtler as it shows how plastic the phenotype can be, and how much functional developmental reorganisation can be achieved without genetic basis. The two-legged goat was a viable organism with a peculiar set of novel phenotypic features. Interestingly one of these features (the nature of the pelvic bones) is reminiscent of the morphology of kangaroos. No wonder that West-Eberhard has been able to extrapolate an interesting provocative speculation on such basis. Consider the popular depiction of human evolution: the passage from ape to human with the gradual acquisition of the bipedal posture. Do we really need many rounds of natural selection, many generations of gradual genetic change in order to get this far? After all, nature required just one trial to create the bipedal goat, that is, an unusual but functional and viable phenotype.

Going back to the crucial issue of inheritance of developmentally generated phenotypes, evo-devo research is producing theoretical advances that will hopefully soon be matched by experimental ones. West-Eberhard believes that environmental induction is the answer to the puzzle: if the environment induces similar changes in the germ then there is no puzzle; otherwise, if the environment induces changes in the soma then we go back into a Baldwinian-plus-genetic-accomodation scenario. Most importantly, what West-Eberhard [22] stresses with vigour is that environmental induction can be as reliable as mutation and that it can even make mutation unnecessary:

> A mutant gene may seem more dependable, that is, more likely to persist across generations, than a novel environmental factor. However, it takes only a little reflection on the nature of development to appreciate the dependability of environmental factors. All organisms depend on the cross-generational presence of large numbers of highly specific environmental inputs: particular foods, vitamins, hosts, symbionts, parental behaviours, and specific regimes of temperature, humidity, oxygen, or light. Such environmental elements are as essential and as dependably present as are particular genes; some, such as photoperiod and atmospheric elements like oxygen and carbon dioxide, are more dependably present than any gene in particular habitats and zones, so we forget that these environmental factors constitute powerful inducers and essential raw materials whose geographically variable states can induce developmental novelties as populations colonize new areas .... Given the evidence, familiar to everyone, that numerous environmental inputs are consistently supplied (essential) during normal development, the scepticism of biologists regarding the reliability of environmental factors relative to that of genes has to rank among the oddest blind spots of biological thought.

Furthermore, from a macroevolutionary point of view, environmental induction solves one of the biggest problems affecting neo-Darwinism: environmental induction can affect many organisms at the same time, and even entire populations, depending on the environmental factor under discussion. The consequences of such process for speciation can be tremendous, especially given stressful conditions [23].

A recent study [24] investigating the developmental mechanisms governing segment formation in arthropods serves to illustrate the promise of the concepts of phenotypic plasticity and environmental induction. The study focuses on a species of centipedes: *Strigamia maritima*. Arthropods are interesting because they pose a puzzle for purely Darwinian accounts of evolution. In fact, most arthropod species are invariant in segment number. It is therefore difficult to understand how, given the lack of within-species variation, selection can "generate" diversity in segment number among this group of organism. The paper illustrates evidence in favour of the hypothesis that segment number formation is influenced by the direct environmental effect of temperature during embryogenesis: higher temperature produces more segments. The authors argue that phenotypic plasticity explains why many species of arthropods tend to exhibit more segments in warmer regions. Note that, first, such study is focused on phenotypic evolution and that genetic change follows phenotypic change: it is up to environmental influence and ensuing phenotypic plasticity to create the intra-specific variation in segment number on which selection can act. Secondly, note that it is environmental induction and not mutation that generates novel patterns of segmentation. The most general extrapolation of the study is that speciation patterns in arthropods could be partly due to simultaneous induction of specific plastic responses on many members of the population.

What I have given here are just hints of the theoretical complexity and variety of evolutionary developmental biology research. Summing up, there are a few lessons we can take from evo-devo. More generally, we could say that evo-devo research has shown that development is evolutionarily relevant in a variety of ways. In this brief section I have emphasised the contribution of approaches to evo-devo that make two aspects of developmental processes particularly clear. First, that there exist a variety of modes of evolution depending on the nature of the available variation: without knowing the details of the mechanisms governing the generation of genetic as well as phenotypic variation, our knowledge of the evolutionary process is impaired. The implication of this perspective is an attack on gradualism, as Darwin intended it: variation is not isotropic but rich and structured. Variation-generation processes work under constraints that limit but also direct the generation of potential variants. The alternative proposed are somehow saltational: hence *natura facit saltum of some kind*, as some evo-devo research vindicates [12, 25], and as comparative genomics is making increasingly clear (via genome duplication and subsequent re-functionalization, and especially via whole-genome duplication events). In any case, what evo-devo wants to make clear is that *nothing in variation makes sense except in the light of development* (a slogan coined by the developmental biologist Jernvall).

A second prominent strand of evo-devo research on which I focused contends that *genes are followers in evolution* (an expression created by West-Eberhard) and that developmental plasticity is paramount. Here the attack on neo-Darwinism concerns the downplaying of the role of genetic mutation. I tentatively tried to explore aspects of the huge and tremendously interesting literature on phenotypic approaches to evolution that draw on phenomena of phenotypic plasticity, developmental reorganisation, and the inducing role of the environment. The biological and philosophical consequences of such approaches are varied, complex and profound. For instance, the focus on phenotypic plasticity and the inducing role of the environment as a counterbalancing presence to the influence of genetic factors in evolution impels a radical change of perspective: the environment, however defined, assumes a fundamental importance that is primary and not anymore subsidiary when compared to that of DNA sequences. The *rediscovery of the environment*, of context in more abstract terms, is one of the fundamental results of contemporary biological research that transcends but deeply influences evo-devo research.

Wonders of the Microcosm

In this section I will focus on bacteria and their extraordinary capacities to respond to environmental challenges. Bacteria and archaea dominate this planet: they have done so and will arguably always do so. The diversity of microbial life is metabolically extremely varied. In fact microbes are ubiquitous and can be found almost everywhere on the planet, and possibly beyond (i.e. consider the panspermia hypotheses). Especially in the philosophy of biology these incontrovertible facts are not given sufficient attention. Multicellularity emerged because bacteria created the conditions for complex multicellular life. Among biological events that can be classified as major evolutionary transitions, the emergence of the ability to perform oxygenic photosynthesis that has changed the atmosphere on our planet is arguably the most important [26]: in a sense, the humble cyanobacterium (aka blue-green algae) is ultimately responsible for a cascade of evolutionary events that eventually led to us.

Even though the philosophy of microbiology is still in its infancy, bacteriology has always been a pivotal part of biology. The work of Avery, Luria, Delbrück and the Lederbergs, just to refer to a few eminent figures in last century's history of biology, was heavily reliant on bacterial model organisms. Bacterial genetics and molecular microbiology produced some of the momentous insights in the history of the discipline and continue to do so. Furthermore, microbiology has also assumed fundamental relevance in evolutionary biology: in fact, some of its results have informed generations of evolutionary biology students and provided heuristic principles directing research. One of such examples, dating back to 1943, is the famous fluctuation test, which sanctioned the official neo-Darwinian stance on bacterial mutation. It says that bacterial mutation is not a Lamarckian response to need; mutants arise at a constant rate during growth, independently of any selection pressure and environmental influence (e.g. physiological stress). The demonstration that even "odd" bacteria – considered by some evolutionary biologists of the time as an exception among "normal" organisms (i.e. animals and plants) – were evolving in a Darwinian fashion gave impetus to the Modern Synthesis. The experiments by Luria and Delbrück and then the Lederbergs in the 1940s and 1950s were considered "crucial" experiments. Philosophers of science might recall that Duhem established, paradoxically once and for all, that there cannot be crucial experiments; but actual scientific practice is different. Even a cursory look at the influential

D. Futuyma's "Evolution" textbook makes one realise that such experiments are presented as the ultimate and definitive demonstration that genuine directed mutation in bacteria cannot happen.

This is a clear historical example of hardening of the Modern Synthesis. The crucial experiments were used in order to discount possible Lamarckian phenomena and processes in one stroke, and in order to trivialise the Lamarckian position. A classic Kuhnian paradigm was build. The assurance that the Darwinian notion of the randomness of mutation had been vindicated across the whole spectrum of organisms studied meant that genuine experimental anomalies were just put under the carpet. But the experimental limits of the fluctuation test were well known from the start. In fact, Delbrück himself admitted in 1946 that the fluctuation test had limited scope, as it could not rule out the existence of adaptive mechanisms of mutation, the reason being that the selective pressure applied to bacteria in the test was too strong. Experimental anomalies contradicting the received view accumulated in the 40 years or so that climaxed with a timely paper by Cairns et al. [27]. The year 1988 also coincides with the resurgence of philosophical interest in supposedly uninteresting bacteria.

Cairns et al.'s [27] paper brought the Lamarckian idea of directed mutation back from limbo. In the last 20 years the phenomenon of bacterial mutagenesis has been intensely studied [28]. The phenomenon has assumed a variety of names (the favoured one is "adaptive mutagenesis") and it is generally recognised that it is complex in involving many mechanisms and processes. It is generally agreed that the phenomenon is compatible with the Darwinian central tenet, namely that mutations are not more likely to be beneficial than not [29]. However, all the other auxiliary hypotheses of the neo-Darwinian paradigm have been abandoned. In fact, contrary to what neo-Darwinians originally thought, bacterial mutations can be environmentally induced, are not solely due to replication errors, can target specific parts of the genome, and are not merely due to the breakdown of the cellular machinery of DNA repair. So, even though the partial "blindness" of bacterial mutational responses to need has been substantially vindicated (how could it be otherwise unless we postulate some kind of mysterious foresight on the part of the bacterium?), all the other tenets of the neo-Darwinian view have been rejected. What can safely be said is that there has been a "softening" of the neo-Darwinian perspective in many respects [30]. The softening is so extensive that Jablonka and Lamb [31] ask whether to call the emerging view on adaptive mutagenesis "neo-Darwinian" instead of "Lamarckian" makes biological sense at all.

Research on adaptive mutagenesis teaches a series of lessons that transcend the Darwinian-Lamarckian debate. What is increasingly clear is that bacteria are able to produce a variety of intelligent responses to the challenges of the environment (what Jim Shapiro [32] calls *natural genetic engineering* capacities). Two aspects of this research should be highlighted. First, contemporary microbial research seems to be more interested in studying organisms in the wild. This has been possible because of the refinement of genetic techniques. It is a very important step towards the development of a powerful science of environmental microbiology. The environmental approach is necessary because it has been estimated that

99% of microbial strains are not culturable. Furthermore, it has been estimated that micro-organisms living in their natural conditions (like biofilms and other multi-taxa communities) are much more resistant to antibiotics than their isolated ken. One of the reasons for this is that bacteria exchange genes: for instance, when stressful conditions activate the SOS response and the cellular repair mechanisms are as a consequence impaired, *E. coli* accepts DNA from *Salmonella enterica* in order to relieve the stress. Research on wild strains of *E. coli* [33] showed that the phenotypic responses of such strains are much more varied, targeted and stress-specific than those of artificial strains. Our ignorance of the evolutionary dynamics of the microcosm is so deep that whenever we scratch its surface we discover amazing and unexpected new phenotypes and processes. Philosophical interest on microbes should centre on such emerging knowledge and themes.

It is only from very recently that bacteria are starting to generate serious rather than fringe philosophical interest. There is no space here to speculate on the reasons for such neglect. The human mind is certainly biased towards the visible, even though bacteria create complex visible aggregates (i.e. biofilms). Multi-cellular organisms are assumed to be more complex and therefore more interesting from an evolutionary point of view. What this otherwise justifiable perspective misses is the simple fact that multi-cellular organisms are ecological communities. It is striking to acknowledge that the human body is constituted only for one tenth of human cells (assumed to share one single genome), and for the rest of prokaryotic symbionts (without considering omnipresent latent viruses such as *herpex simplex* or cold producing *rhinoviruses* – see next section). The notion of *microbiome* (coined by J. Lederberg to indicate the vagaries of symbionts cooperating with organismal cells in the functioning of multi-cellular organismal physiology) is increasingly seen as relevant because, for instance, many human physiological processes consist in the coordination and interaction of the activities of human cells and bacteria. Aspects of the physiology of human digestion (e.g. fat storage) are a classic case, so much so that there exists a lucrative bio-industry profiting from selling pro-biotic products that are supposed to correct our gut imbalances (e.g. bacillus reuteri). Consumption of yogurt across the Middle East and India is indicative of a less consumeristic popular wisdom. The philosophical irrelevance of bacteria becomes even more puzzling if we also take into account the incredibly important role bacteria (and more generally micro-organisms) play in the biotechnology industry (from the food to the pharmaceutical to the bioremediation industry).

It has surely not escaped notice that evo-devo is primarily about creatures with embryos, whether they are animals or plants. Evo-devo provides an evolutionary perspective on visible organisms (especially metazoa) and is not concerned about the vast majority of unicellular organisms inhabiting our planet: other eukaryotes, but most of all bacteria and archaea. But this focus, though legitimate in some respects, fails to take into account the real nature of life: its complex interactive dynamics. There are some approaches in evo-devo that take seriously into account the amount of integration between different kinds of organisms. Scott Gilbert [34] called this new integrative approach "ecological developmental biology". In the course of the last decade the spectrum of ecological and epigenetic factors that has been taken into consideration for the proper understanding of developmental processes has widened considerably: while the initial emphasis (I presume mainly for reasons related to the availability of data) was on environmental factors such as temperature, PH and population density, a variety of new studies stresses the developmental contribution of bacterial organisms and viruses. The latter contribution has also undergone evolution: while 10 years ago bacterial and viral contribution to development was mainly considered under the rubric of infectious agents, these days researchers have understood that, given the way multi-cellular organisms are organised, bacteria play a fundamental role in producing viable developmental outcomes. West-Eberhard [22] reports that studies on zebrafish show that environmentally supplied bacteria regulate the expression of 212 genes: "Without these bacteria, many of which produce highly specific host responses, the developing fish die."

Ecological developmental biology is one of the many expressions of a research Zeitgeist that aims to capture the holistic character of biological processes. In this sense it could be seen as an instance of a more general movement labelled "integrative biology" that promises to incorporate various biological disciplines that originally developed independently but that in the light of new molecular findings and comparative genomics studies are becoming more easily interlinked.

Focus on the interaction between microbiome and organismal cells has a variety of philosophical implications. If bacteria and viruses are not seen as merely infectious, and if developmental processes are partially regulated by such biological entities, then our conceptions of disease will radically change. Traditional philosophical issues concerning personal identity will also radically change if the notion of microbiome is seriously taken into account: are we what we are just because our human cells all possess our unique genome (without taking into account cases of genomic mosaicism), or do we need to consider the enlarged environment, the wider context?

Focussing again on evolutionary themes, what the conception of microbiome makes clear is that the amount of co-operation between organisms belonging to different species and domains has been hugely underestimated. Gilbert and Epel [35] refer to many of such interactions as *developmental symbioses*. The symbiosis between human cells and *e. coli* in the human gut is perhaps the most well studied example: *e. coli* bacteria are developmentally, immunologically and physiologically essential for our health. In this case, Gilbert and Epel conjecture that the existence of many instances of developmental symbiosis strengthen the idea that group selection was a powerful force in evolution, as already convincingly argued by Sober and Wilson [36]. From another standpoint, Dupre' et al. [37] argue that a metagenomic perspective should be endorsed, one that considers the metagenome (all the DNA present in a community) as a community resource, while constituent individual organisms should be seen as ontological abstractions.

More generally, I believe that the science of genomics has many potential ontological implications. First, it is making clear that certain developmental phenomena like mosaicism and chimerism are very common, even possibly endemic. In this respect the idea of individual organism as being defined in terms of one lineage of genomically identical cells or possessing one pure and unique germ-plasm is biologically problematic. The argument could be stretched as far as sustaining that all organisms are multi-lineal and multi-genome. This is valid both for unicellular bacteria and archaea, who happily exchange genetic resources, and for multi-cellular eukaryotes, which should probably be better seen as constituted of a variety of assembling parts with different genetic origins. This view of macro-organisms or "macrobes" [43], as multi-genome as multi-genome and multi-lineage genetic associations is particularly at odds with our intuitions about individuality. One could venture to argue, stretching Gilbert and Epel's defense of group selection, that genomics poses new sets of issues concerning the units of selection debate: even if it is undoubtable that genes and organisms are units of selection, that groups and species have a good chance of being recognised as such, it remains clear that, given the way selection works (i.e. at many levels of biological organisation), even the emerging associations genomics and microbiology are disclosing (e.g. biofilms, units consisted of organisms and parasitic mobile genetic elements such as plasmids, tight multi-lineal and multi-genomic communities of organisms belonging to different taxa such as consortia etc.) could fulfill the role of unit of selection [38]. The selective role of associations is, of course, not new. Maynard-Smith [39] speculated that in symbiogenesis natural selection does not behave in a Darwinian way (i.e. at the level of the individual organism or gene), but that it acts, so to speak, for the benefit of the host. But this view is at odds with the data showing that selection can act at the level of the partnership, as indicated, for instance, by the strict associations formed by amoebas and bacteria [40]. In any case, a nature filled with communal gene pools, cooperative interactions across species, and symbiotic relationships is very different from the ultra-competitive world popularised by ultra-Darwinians like Dawkins. Rather than being red in tooth and claw, nature is immensely co-operative.

The social dimension of microbial life is under intense research scrutiny these days. Some biologists believe that we can effectively talk of an emerging new science: sociobacteriology [41]. Processes like biofilm formation, quorum sensing and chemotaxis are increasingly understood. All these processes can be described as including elements of learning: bacteria communicate by sensing environmental variables or by releasing molecules. That bacteria were capable of communicating and exchanging genetic resources was already discovered by Avery and collaborators back in 1944. Avery's team wanted to find out how harmless pneumococci became infectious by coming into contact with dead but poisonous pneumococci. He discovered that, by digesting pieces of DNA scattered around the environment, harmless bacteria learned how to become deadly. Incidentally Avery and colleagues also discovered that DNA was the stuff of genes, the basis of inheritance: in fact, the material transmitted from dead and poisonous to harmless bacteria was DNA. Another significant aspect of the results is that it was discovered that the transmission of DNA information was not only vertical (from parent to offspring) but horizontal/lateral (in all possible directions).

Processes of lateral genetic transfer within species analogous to forms of family communications do no strike us as particularly revolutionary, but similar processes between individuals of different species are rather unsettling to many evolutionary biologists [42]). While evolutionary biologists accept that lateral gene transfer is common in the microworld, most would deny that it affects complex eukaryotes. One fundamental reason at the basis of this denial is that the Darwinian conception of the tree of life with all its neat ramifications would be compromised if lateral gene transfer were a common process. Genomics is, as a matter of fact, vindicating this stance: the most accurate depiction of the interrelationships between life forms is not a tree but a network (I will return to this issue in the next section too).

Wolbachia pipientis is a parasitic bacterium that has now become an evolutionary celebrity. Until 10 years ago it was thought to parasitise some species of insects but nobody could predict its pervasiveness. The relationship between Wolbachia and its eukaryotic hosts is nowadays intensely studied. A recent genomic study [43] shows that LGT happened between Wolbachia and many of its parasitized insects (as well as other studied hosts), vindicating the hypothesis that lateral gene transfer can occur between bacteria and eukaryotes. The consequences for speciation are also potentially interesting. If lateral gene transfer is indeed a process that allows genetic exchange between species even belonging to different domains, then it provides the means for the acquisition of novel phenotypes. As already highlighted by West-Eberhard [22], the geographical separation of species accompanied by the multigenerational process of genetic diversification is not necessary for speciation: what suffices is a humble bacterium like Wolbachia (or an even humbler virus), acting as a genetic shuttle by inserting the potential for novel phenotypes in a host subpopulation. In this case speciation could be very quick.

To conclude, the challenges to mainstream evolutionary thinking coming from a focus on bacterial evolution are many. In this section I have briefly reviewed research showing that the cooperative nature of life means that the organisms constituting developmental symbiotic relationships could be seen as natural units of evolution favoured by group selection. I have also tried to briefly show that lateral gene transfer processes could potentially re-write the history of prokaryotic as well as eukaryotic life. But there is a theme that deserves particular attention. There is some common ground between environmental microbiology and ecological developmental biology: organisms at all level of biological organisation show incredible phenotypic plasticity. In the first part of the section I tried to give hints concerning the way in which bacteria, particularly in the wild, creatively react to environmental stresses: such capacity is, in my opinion, another instance of the universal phenomenon of phenotypic plasticity.

Virolution

If there is a branch of biology that has always been neglected by mainstream evolutionary theory that is virology. Disquisitions abound even concerning the nature of viruses: are they alive at all? If they are not, why should we care about them? According to Luis Villareal the two questions are interconnected: in fact, if one thinks that viruses are not alive, that they simply are protein-coated pieces of host genetic material, then one is bound to downplay their evolutionary role [44].

The question concerning the ontological status of viruses is fuzzy. It is true that viruses are parasites needing host cells to replicate, but so do the constituents of cells: it is an odd question to ask whether ribosomes and mitochondria are alive. It is certainly true that viruses are not autonomously capable of performing complex biochemical tasks; however, they somehow resemble seeds in having the "potential" to spring to life, given the right environment. Being constituted of chemical compounds, viruses are capable of many extraordinary enterprises. For instance, they are the only biological entities capable of "self-resuscitation". The phenomenon, called "multiplicity activation", consists in the capacity to re-assemble from parts: presence of viral debris is sufficient for a viral genome to restore its functionality; the trick is to re-assemble from the bits scattered in the cytoplasm. Viruses might not be fully alive (i.e. autonomous), but they certainly differ from inanimate matter in being able to resurrect! In any case, to continue the analogy, given that seeds have profound evolutionary impact, the same must be true of viruses.

I am not claiming that virology has been neglected by biologists. On the contrary, it has always been a thriving part of biology, in part because viruses are creatures easy to work with in laboratory settings. The pioneering work of Nobel Laureates Luria and Delbrück on bacteriophages paved the way to the emergence of many molecular techniques, and many more Nobel Prizes have been awarded for virology research. For example, three Nobel Prizes have been awarded for the discovery of reverse transcriptase, a process that was somehow denied at first because supposedly contravening the central dogma of molecular biology. But the general perspective of the evolutionary biology community concerning the evolutionary role of viruses has always been constrained by the assumptions that viruses are to be considered as either mere carriers of disease, or as vaguely interesting parasites, or more sympathetically as vectors for the insertion of alien genetic material on the host genome (a process called transduction).

It is in this latter sense that viruses have made an evolutionary return of sorts in the last 20 years or so. Some people might remember Ted Steele's Lamarckian claims concerning the inheritance of acquired characters via a mechanism of reverse transcription involving retroviruses inserting somatic information back onto the germ line. Steele, ostracised by peers, even tried to ask Sir Karl Popper – someone who had written extensively against Lamarckism, but who nonetheless kept an open mind – for help (this very interesting story is told by Aronova [45]). The conception of viruses as vectors of novel genes is not dissimilar from the conception of bacteria as vectors seen in the last section with Wolbachia: the evolutionary effects of parasitic infections are straightforward in principle, as both provide suitable mechanisms to cross the Weismann barrier.

However, it was only with the advent of genomics that the evolutionary fortunes of viruses started to change drastically. Genomic techniques are destabilising some of biology's most stable and cherished beliefs about phylogenetic relationships among organisms. In fact, what is being discovered is that many viruses have posed their signature on the genomes of many species, including us. Placenta formation in all (placental) mammals is developmentally regulated by genetic elements of viral origin [46]. Many parts of the human genome have viral origin (especially in the form of endogenous retroviruses, aka HERVs). While most of these sequences seem to be not expressed, some are. Among these latter some are involved in the development of the trophectoderm of the placenta, a very significant and complex part of our reproductive organ. This case of retroviral symbiosis is also an example of virus-mediated developmental symbiosis in Gilbert et al. [33] sense. Villareal [44] argues that this means that our human identity is clearly partially due to viruses.

One of the strongest reasons for legitimising the evolutionary interest in viruses springs indirectly from the replicative power that is at the basis of their creativity: generation time for viruses is very short; if one adds that viral populations are extremely large and that viruses are possibly even more ubiquitous than micro-organisms, then it is intriguing to speculate that viruses might be tremendously creative, that they have the potential to create novel genes at a much higher rate than cellular organisms. The extent of viral genetic creativity is unsurpassed. It is also largely unknown, though not surprising. What is known is in any case very promising. Villareal et al. [47] showed that, contrary to what is generally believed – namely that viruses are genetic debris – 80% of viral genes have no counterparts in the eukaryotic genetic database, which clearly contradicts the favoured hypothesis that viruses originated from cellular genomes. Villareal [48] also reports that genomic evidence shows that many basal versions of genes are often viral in origin, including the most basic genes controlling the processes of DNA replication in eukaryotes. Patrick Forterre has speculated that even DNA itself has viral origin [49]. New studies reveal that viral footprints can be retraced in the creation of many regulatory and epigenetic pathways, in the emergence of key DNA polymerases in eukaryotes as well as in the origin of introns [50]. Viral metagenomic research shows that viral diversity is enormous even *vis a vis* prokaryotes. Several metagenomic studies seem to show that at least in some environments (e.g. marine habitats) bacteriophages are the most abundant biological entities by an order of magnitude [51]. It would be easy to dismiss these results on the basis of the smallness of viral genomes compared to cellular ones, but this would amount to deny their obvious evolutionary significance. In particular, viruses with large genomes possess a substantial part of genes with no homologs in current sequence databases (Koonin [51]. To put it in very cautious words, this seems to imply that viral genomes comprise a large part of the genetic diversity on our planet.

Particularly striking in this context seems to be the evolutionary contribution of retroviral symbioses. In a series of papers Ryan [50, 52] has recently argued that viral symbioses, – technically viral infections of the germ cells of the host – are increasingly recognized as being of paramount importance from a medical point of view. When viruses enter the host and "endogenise" (i.e. enter the germ line and start to be transmitted in Mendelian fashion) – a process by no means universal but frequent – they constitute with a host a "holobiont", i.e. a symbiotic union of virus and host with a single genome. Endogenous retroviruses are known, especially in the human case (where they are being intensely studied for medical reasons), to control genes expression of host genes and to play a variety of physiological and developmental roles. What is most intriguing from an evolutionary point of view is the fact that despite the process of endogenization – which strongly limits the infective power of the virus by transforming it in a persistent and eventually asymptomatic infection – the genetic contribution of viral origin is not totally silenced but remains partly active. Furthermore, there is increasing evidence that endogenous retroviruses retain the ability to create new genes [50]. Once endogenous retroviruses enter the host genome, they partially retain their ancestral capacity for evolutionary innovation intact despite concomitant evident degradation. Some active sequences retain the capability to interact with new endogenous retroviruses or other genetic components. Thus, integration in the genome does not equate to become "junk DNA". This could mean that the process of "infection" should be seen in a new light as constituting one of the ways in which retroviral symbiosis plays a role in the emergence and evolution of the capacity of the genome to react to environmental change (i.e. genomic plasticity [53].

A further paramount reason for considering viruses as evolutionary significant pertains to their ability to constitute an interconnected pool of highly transferable genetic resources. The conception of "mobilome" or "virosphere" must, to be sure, not be confined to viruses as it includes a variety of sub-cellular biological entities, such as plasmids for instance, that can be labeled "selfish replicons" because of their incapacity to perform translation independently. To view viruses and plasmids as merely "parasitic" entities reiterates a conception of sub-cellular life connected with disease that is not evolutionary pertinent. Even though plasmids are the *culprits* from our human perspective – for instance for being the vectors that provide the means to bacteria in order to elude our antibiotic weapons – they are from the perspective of the bacterium an important resource of useful phenotypes. Plasmids, viruses and other so-called "mobile genetic elements" are increasingly at the centre of biological investigation. The principal reason is that comparative genomics and metagenomics vindicate the hypothesis that mobile genetic elements make a crucial contribution to cellular genomic structure. While this hypothesis is totally vindicated as far as prokaryotes are concerned, it remains to be seen how it fares with eukaryotes. What can be said at this stage is that the transfer of segments of bacterial genomes mediated by mobile genetic elements to the genome of eukaryotic (especially unicellular) hosts is common [51].

The ways in which mobile genetic elements shape cellular genomes are multifarious. Our ignorance on the matter is deep but the clear importance of the process is changing our research attitudes [54]. The fundamental point is that plasmids and viruses travel across species, domains and lineages by making important genomic contributions. They are, in brief, a crucial vehicle of integration of genetic and phenotypic resources. There are varieties of ways in which such integration can take place across the biosphere. Lateral gene transfer occurs through the horizontal acquisition of novel functions (mostly enhancing adaptability and robustness) via genetic means mediated by mobile genetic elements with subsequent genome integration. But lateral gene transfer is only one way in which biological resources can be integrated. Another fundamental process is endosymbiosis, which occurs via the horizontal incorporation of sub-cellular and cellular biological entities in the cytoplasm and possibly genome. While lateral gene transfer processes necessitate genome integration, endosymbiotic ones do not. But integration of resources through symbiotic processes is even wider in means and scope, ranging from cases of vertical acquisition of entire organisms without genetic transfer, to cases of vertical infection through the germplasm (e.g. ERVs), to cases of exosymbiosis through the environmental acquisition of adaptive symbiont layers playing crucial physiological and development roles for the host (e.g. the bacterial flora in our gut). Symbiotic relationships are very varied and can be categorized in multiple ways by taking into account the nature of the information exchanged (e.g. genetic or metabolic), the nature of the organisms involved (e.g. eukarya, bacteria, cellular or even sub-cellular biological entities), the existence of physical contact between host and symbiont, the existence of gene transfer and genetic integration etc. New data vindicate the hypothesis that "cross-lineage borrowing" and resource transfer is ubiquitous in nature. The evolutionary relevance of lateral gene transfer and symbiogenesis will most surely assume increased importance the more we study associations of organisms.

Let us move to considering the subject of the evolutionary significance of these findings. Villareal [44, 48] is strongly supporting a viro-centric view of evolution that would downplay the contribution of processes of genetic mutation followed by selection. Such view is based on two basic elements: first, on the unsurpassable power of exploration of chemical possibilities and genetic design space attributable to viruses, which makes them likely leading producers of evolutionary novelty; and secondly on the knowledge of mechanisms of DNA insertions in the cellular genomes. Viruses, as already noted by Luria, offer a new system of inheritance and transmission of the genetic novelty they produce by effortlessly merging to and re-emerging from the cellular genome. Only future genomic research will show how far this viro-centric view of evolution is correct, and how much it challenges neo-Darwinian orthodoxy. What can be easily argued is that the existence of lateral gene transfer and of symbiotic means of integration makes the species, lineage and domain boundaries permeable. Lateral gene transfer and symbiosis show that it is not necessary that all improvements leading to evolutionary change and novelty must happen *within* a lineage. As a consequence, it seems safe to state that genetic innovation comes from mutation as well as from integrative processes such as symbiosis and lateral gene transfer. The consequences for traditional tree-based approaches to phylogenetics are intensely under scrutiny. In fact, it seems that the Darwinian simile of the Tree of Life is problematic in light of the data yielded by comparative genomics [55]. Nonetheless, in all these cases it must be highlighted that natural selection retains a fundamental role: even though it does not *create* the relationship, it *edits* the new partnership.

Conclusion

Does Darwinism remain important? In order to give an articulated answer to this question we need to consider two aspects. First of all, we can ask whether biology has provided us with suitable alternatives to selection. I think the answer to this question is complex and cannot be attempted in this context. A second aspect of the issue concerns the necessity and nature of the revisions to neo-Darwinism. What I have said in the last three sections is in striking contrast to what Mayr said in the quote at the start of the paper. But the contrast is partly spurious. Consider how Mayr [2] continues his argument:

> But there is something else that has indeed affected our understanding of the living world: that is its immense diversity. Most of the enormous variation of kinds of organisms has so far been totally ignored by the students of speciation. We have studied the origin of new species in birds, mammals, and certain genera of fishes, lepidopterans, and molluscs, and speciation has been observed to be allopatric (geographical) in most of the studied groups. Admittedly, there have been a few exceptions, particularly in certain families, but no exceptions have been found in birds and mammals where we find good biological species, and speciation in these groups is always allopatric. However, numerous other modes of speciation have also been discovered that are unorthodox in that they differ from allopatric speciation in various ways. Among these other modes are sympatric speciation, speciation by hybridization, by polyploidy and other chromosome rearrangements, by lateral gene transfer, and by symbiogenesis. Some of these nonallopatric modes are quite frequent in certain genera of cold-blooded vertebrates, but they may be only the tip of the iceberg. There are all the other phyla of multicellular eukaryotes, the speciation of most of them still quite unexplored. This is even truer for the 70-plus phyla of unicellular protists and for the prokaryotes. There are whole new worlds to be discovered with, perhaps, new modes of speciation among the forthcoming discoveries.

I interpret Mayr's words as a kind of implicit denial of his previous words. Here Mayr is admitting that our ignorance is profound, that known biological mechanisms of speciation will assume new relevance in the future, and possibly that new ones will be discovered. This is what I have been trying to show in the three previous sections. The knowledge stemming from future biology will necessarily shape our understanding and interpretation of Darwin's legacy and thinking.

I would like to conclude by making two points. The first is by stating that Darwin's idea of natural selection is absolutely central and fundamental in biology, whether you see it as temporally primary, directional, positive and creative or as merely purifying, subsidiary and stabilising. The idea is so important that it is arguably universal. It is universal because it applies to all life domains and to all biological entities. The same cannot be said of many domain-specific mechanisms on which specialised practitioners in many areas of biology focus, with some exceptions (e.g. phenotypic plasticity, a very pervasive feature of organisms at all levels). Dawkins' idea of Universal Darwinism (originally proposed by Donald Campbell, and in some limited form by Popper) seems to be deeply true. Recent research [56] shows that prions (i.e. biological entities devoid of genetic material) can evolve in a Darwinian fashion. The authors of the study suggest that Darwinian evolution does not require DNA or RNA, as it works on any kind of substrate. Campbell [57] and Popper [58] gave similar examples by referring to crystal formation and chemical evolution.

I would also like to make a further point. Evolution is not only a story about selection. Evolution is a complex process regulated by the interaction between selection, neutral (e.g. drift) and compositional processes, where with the latter expression I refer to phenomena of integration of resources based on the structured (e.g. modular) organisation of evolutionary units. Examples of such processes are the various phenomena of exchange of genetic, metabolic and other phenotypic resources across cellular lineages (e.g. lateral gene transfer based on mobile genetic elements). Some eminent contemporary biologists (Lynch [59], Koonin [51]) speculate that future evolutionary biology will be more about tinkering and integration and less about adaptation, and that the picture of evolution emerging from contemporary biology will be quite different from the neo-Darwinian one.

References

1. Gould, S.J.: Bully for Brontosaurus. W.W. Norton & Co., New York/London (1991)
2. Fracchia, J., Lewontin, R.C.: Does culture evolve? Hist. Theory **38**, 52–78 (1999)
3. Mayr, E.: Happy birthday: 80 years of watching the evolutionary scenery. Science **305**(680), 46–47 (2004)
4. Coyne, J.A.: Why Evolution Is True. Viking, New York (2009)
5. Dawkins, R.: The Blind Watchmaker. Longmans, London (1986)
6. Gould, S.J.: The Structure of Evolutionary Theory. Belknap, Harvard (2002)
7. Minelli, A.: Evolutionary developmental biology does not offer a significant challenge to the Neo-Darwinian paradigm. In: Ayala, F., Arp, R. (eds.) Contemporary Debates in Philosophy of Biology. Wiley-Blackwell, Oxford (2009)
8. Müller, G.B.: Evo-devo: extending the evolutionary synthesis. Nat. Rev. Genet. **6**, 1–7 (2007)
9. Gould, S.J.: Is a new and general theory of evolution emerging? Paleobiology **6**, 119–130 (1980)
10. Hoekstra, H.E., Coyne, J.A.: The locus of evolution: evo devo and the genetics of adaptation. Evolution **61**(5), 995–1016 (2007)
11. Carroll, S.B.: Endless forms: the evolution of gene regulation and morphological diversity. Cell **101**, 577–580 (2005)
12. Minelli, A., Chagas-Júnior, A., Edgecombe, G.D.: Saltational evolution of trunk segment number in centipedes. Evol. Dev. **11**(3), 318–322 (2009)
13. Dawkins, R.: The Extended Phenotype. Oxford University Press, Oxford (1982)
14. Sober, E.: Philosophy of Biology. Westview Press, Boulder (2000)
15. Minelli, A.: Forms of Becoming. Princeton University Press, Princeton (2008)
16. Amundson, R.: Two concepts of constraint: adaptationism and the challenge from developmental biology. Philos. Sci. **61**, 556–78 (1994)
17. Gilbert, S.F.: The generation of novelty: the province of developmental biology. Biol. Theory **1**(2), 209–212 (2006)
18. Ruse, M.: Does evoDevo break the paradigm? Biol. Theory **2**(2), 182 (2007)
19. Popper, K.R.: Objective Knowledge. Oxford University Press, Oxford (1972)
20. West-Eberhard, M.J.: Developmental Plasticity and Evolution. Oxford University Press, Oxford (2003)
21. Kirschner, M., Gerhart, J.C.: The Plausibility of Life. Yale University Press, New Haven (2005)
22. West-Eberhard, M.J.: Developmental plasticity and the origin of species differences. PNAS **102**, 6543–6549 (2005)
23. Badyaev, A.: Evolutionary significance of phenotypic accommodation in novel environments: an empirical test of the Baldwin effect. Philos. Trans. R. Soc. B. **364**, 1125–1141 (2009)

24. Vedel, V., Chipman, A.D., Akham, M., Arthur, W.: Temperature-dependent plasticity of segment number in an arthropod species: the centipede Strigamia maritima. Evol. Dev. **10** (4), 487–482 (2008)
25. Cebra-Thomas, J., Tan, F., Sistla, S., Estes, E., Bender, G., Kim, C., Gilbert, S.F.: How the turtle forms its shell. J. Exp. Zool. **304B**, 558–569 (2005)
26. Knoll, A.H.: Life on a Young Planet: The First Three Billion Years of Evolution on Earth. Princeton University Press, Princeton (2003)
27. Cairns, J., Overbaugh, J., Miller, S.: The origin of mutants. Nature **335**, 142 (1988)
28. Foster, P.: Adaptive mutation in *Escherichia coli*. J. Bacteriol. **186**(15), 4846–4852 (2004)
29. Lenski, R., Mittler, J.E.: The directed mutation controversy and Neo-Darwinism. Science **259**, 188–194 (1993)
30. Brisson, D.: The directed mutation controversy in an evolutionary context. Crit. Rev. Microbiol. **29**, 25–35 (2003)
31. Jablonka, E., Lamb, M.: Evolution in Four Dimensions. MIT Press, Cambridge (2006)
32. Shapiro, J.A.: A 21st century view of evolution: genome system architecture, repetitive DNA, and natural genetic engineering. Gene **345**, 91–100 (2005)
33. Bjedov, I., Tenaillon, O., Gérard, B., Souza, V., Denamur, E., Radman, M., Taddei, F., Matic, I.: Stress-induced mutagenesis in bacteria. Science **300**(5624), 1404–1409 (2003)
34. Gilbert, S.F.: Ecological developmental biology: developmental biology meets the real world. Dev. Biol. **233**, 1–12 (2001)
35. Gilbert, S.F., Epel, D.: Ecological Developmental Biology. Sinauer Associates Press, Sunderland (2009)
36. Sober, E., Wilson, D.S.: Unto Others. Harvard University Press, Cambridge (1998)
37. Dupre', J., O'Malley, M.: Metagenomics and biological ontology. Stud. Hist. Philos. Biol. Biomed. Sci. **38**, 834–846 (2007)
38. Bapteste, E., Burian, R.M.: On the need for integrative phylogenomics, and some steps towards its creation. Biol. Philos. **25**, 711–736 (2010)
39. Maynard-Smith, J.: A Darwinian view of symbiosis. In: Margulis, R., Fester, R. (eds.) Symbiosis as a Source of Evolutionary Innovation. MIT Press, Cambridge (1991)
40. Jeon, K.W.: Genetic and physiological interactions in the amoeba-bacteria symbiosis. J. Eukaryot. Microbiol. **51**, 502–508 (2004)
41. Shapiro, J.A.: Bacteria are small but not stupid. Stud. Hist. Philos. Biol. Biomed. Sci. **38**, 807–819 (2007)
42. O'Malley, M.A., Dupre', J.: Size doesn't matter: towards a more inclusive philosophy of biology. Biol. Philos. **22**, 155–191 (2007)
43. Hotopp, J.C., et al.: Widespread lateral gene transfer from intracellular bacteria to multicellular eukaryotes. Science **317**, 1753–6 (2007)
44. Villareal, L.P.: Can viruses make us human? Proc. Am. Philos. Soc. **148**(3), 296–323 (2004)
45. Aronova, E.: Karl Popper and Lamarckism. Biol. Theory **2**(1), 37–51 (2007)
46. Sugimoto, J., Schust, D.J.: Review: human endogenous retroviruses and the placenta. Reprod. Sci. **16**(11), 1023–1033 (2009)
47. Villarreal, L.P., De Philippis, V.R.: A hypothesis for DNA viruses as the origin of eukaryotic replication proteins. J. Virol. **74**, 7079–7084 (2000)
48. Villareal, L.P.: Viruses and the Evolution of Life. American Society of Microbiology Press, Washington, DC (2005)
49. Zimmer, C.: Did DNA come from viruses? Science **312**(5775), 870–872 (2006)
50. Ryan, F.P.: Genomic creativity and natural selection: a modern synthesis. Biol. J. Linn. Soc. **88**, 655–672 (2006)
51. Koonin, E.V.: Darwinian evolution in the light of genomics. Nucleic Acids Res. **37**(4), 1011–1034 (2009)
52. Ryan, F.P.: An alternative approach to medical genetics based on modern evolutionary biology. Part 2: retroviral symbiosis. J. R. Soc. Med. **102**, 324–331 (2009)

53. Loewer, R., Löwer, J., Kurth, R.: The viruses in all of us: characteristics and biological significance of human endogenous retroviruses sequences. PNAS **93**, 5177–5184 (1996)
54. Frost, L.S.: Mobile genetic elements: the agents of open source evolution. Nat. Rev. Microbiol. **3**, 722–732 (2005)
55. O'Malley, M.A., Martin, W., Dupre, J.: The tree of life: introduction to an evolutionary debate. Biol. Philos. **25**, 441–453 (2010)
56. Jiali, L., Browning, S., Mahal, S.P., Oelschlegel, A.M., Weissmann, C.: Darwinian Evolution of Prions in Cell Culture. Science Express Online, Dec 31 (2009)
57. Campbell, D.T.: Unjustified variation and selective retention in scientific discovery. In: Ayala, F.J., Dobzhansky, T. (eds.) Studies in the Philosophy of Biology. Macmillan, New York (1974)
58. Popper, K.R.: Darwinism as a metaphysical research programme. In: Unended Quest, pp. 167–180. Fontana Collins, Glasgow (1976)
59. Lynch, M.: The frailty of adaptive hypotheses for the origins of organismal complexity. PNAS **104**, 8597–8604 (2007)

Implications of Recent Advances in the Understanding of Heritability for Neo-Darwinian Orthodoxy

Martin H. Brinkworth, David Miller, and David Iles

Introduction

Conventional neo-Darwinism holds that genetic variation (mutation) leads to the appearance of phenotypic variants, on which natural selection can act so that those least well adapted eventually become eliminated from the population. This postulate assumes not only that all heritable variation comes from mutations in the DNA, but also that the mutations appear randomly throughout the genome. Certain exceptions to these principles are known, e.g. adaptive mutation and immunoglobulin somatic recombination, and some regions of DNA are more susceptible to mutation than others (hotspots). However, these are highly specific instances of endogenously induced variation and do not alter the general principle that natural selection acts on random variation. Of more importance for evolution is transgenerational variation through the sperm, eggs and their precursors (germline). Recent findings have revealed instances of non-random inheritance of mutation and non-DNA based heritability, and it remains the case that the source of DNA variation is, in an evolutionary context, poorly understood. Whether or not these exceptions have any implications for neo-Darwinism is therefore a valid enquiry.

M.H. Brinkworth (✉)
School of Medical Sciences, University of Bradford, Bradford BD7 1DP, UK
e-mail: M.H.Brinkworth@Bradford.ac.uk

D. Miller
Reproduction and Early Development Group, Leeds Institute of Genetics, Health and Therapeutics, University of Leeds, Leeds, UK

D. Iles
Faculty of Biological Sciences, University of Leeds, Leeds, UK

M. Brinkworth and F. Weinert (eds.), *Evolution 2.0*, The Frontiers Collection,
DOI 10.1007/978-3-642-20496-8_17,

Variation in Inheritance of Mutation

For phenotypic variation to be inherited, the instructions coding for the variants have to be present in the germline. Mutations can be induced in model systems by chemicals (mutagens) including cyclophosphamide, an anticancer drug that also causes mutations in germ cells, particularly those that will develop into sperm. Treatment of rats with low doses over a prolonged period of time produces mutations in the sperm that result in foetal abnormalities in the offspring when the treated rats are mated with untreated females [1]. Intriguingly, examination of the testes of males treated in this way shows a lower level of germ cell death than in untreated males [2]. Cell death is one of the mechanisms by which organisms protect themselves from damage (particularly genetic damage) so if a mutagen causes damage leading to mutation but also inhibits one of the defence mechanisms for eliminating damage, then it will enhance the amount of mutation that can be passed on to the next generation. The key here is the low dose of cyclophosphamide: it is thought that this maximises the chance of damage eventually occurring, while remaining below the threshold for the induction of defence mechanisms, and possibly even inhibiting them as well [3]. Similarly, environmental exposure to mutagens is generally long term and at low levels, so this process could be relevant in an evolutionary context by enhancing heritable mutation rates. Furthermore, cases where the mutagen also inhibits cellular defence mechanisms could lead to the survival of such cells. Thus, mutations for cell survival might be passed on more frequently than others, in other words, non-randomly. However, this type of mutation is more likely to cause pathological outcomes than to generate potentially useful mutations. The children of men who smoke before conception are in fact slightly more likely to develop cancers such as lymphoma or leukaemia [4], which could be the result of such a process.

Epigenetics

Currently, interest in heritable transmission of variant characteristics is focused on epigenetics, which is the study of heritable changes in gene behaviour that do not involve mutation. These include: modification of DNA by the addition of a chemical group to particular molecules in the sequence (methylation); transmission of RNA between generations (paramutation); and the chemical modification of histone proteins (histone modification), which together with DNA, make up chromatin – the substance of chromosomes.

DNA methylation is a commonly used regulatory mechanism that helps cells to control when individual genes are switched on or off. The pattern of methylation therefore differs between different cell types but it can also be influenced by environmental factors such as diet. Recently, there has been public controversy about the role of DNA methylation in the aetiology of obesity. It has been claimed that obese mothers suffering from diabetes or hypertension produce children who also develop these diseases. However, it is erroneous to assume that environmental factors can directly influence evolution in a Lamarkian way. Any association between diet and DNA methylation that may be associated with obesity-related diseases is not inherited by the offspring, but acquired by them while still in the womb. At that stage of their development, they are exposed to the same dietary factors as the mother via her bloodstream, so it is perhaps unsurprising that they develop similar, aberrant DNA methylation patterns. Unless these offspring are female and continue with the same lifestyle as their mothers when they are adult and themselves become pregnant, the effect is likely to disappear in the next generation [5].

In terms of evolution, external conditions affecting DNA methylation status could have a survival advantage, by allowing organisms to adapt in the short-term to sudden changes in their environment. Depending on how long the environmental stress persisted and the generation time of the organism, this period of respite could possibly also allow time for spontaneous, random mutation eventually to provide a permanent solution to the problem.

Paramutation is another phenomenon that breaks Mendel's first law of inheritance. Here, one variant, or allele, of a gene affects the action of the other variant, resulting in a heritable change in phenotype, even if the allele that caused the change is itself not transmitted. The best known example of this occurs in certain types of maize, where individual kernels can show a colour that does not match that of their genotype. It is thought that this phenomenon may be associated with different mechanisms, including DNA methylation. However, in the Kit paramutant mouse, white patches on the feet and tail tip can occur in mice homozygous for the normal (wildtype) coat-colour allele if there was a heterozygote male carrying the white coat-colour allele among their ancestors. It is now known that this is because mutant RNA persists through the generations and affects coat colour even though the mutant allele is not present [6].

The last example of transgenerationally-inherited epigenetic phenomena is that of the histone code. As mentioned earlier, histones are the proteins that are bound to DNA to form chromatin. During the production of sperm, most of the histones are removed and replaced by different proteins known as protamines. These are much smaller molecules and bind the DNA in a different way, which is much more space efficient and allows the nucleus inside the sperm head to be much smaller than a nucleus from any other cell in the body, thus allowing the sperm to be motile and a nucleus much more efficient swimmer. However, some of the histones are always retained and it has recently been discovered, by ourselves and others, that they are principally retained at parts of the DNA associated with the regulation of gene expression, especially for those genes needed in embryogenesis. In other words, it appears that the retention of the histones marks the regions of DNA that need to be expressed immediately or very shortly after fertilization [7, 8]. This could overturn the dogma that the egg provides all the materials and gene products that the embryo needs for the early stage of its life. But of more relevance here is the fact that it could explain why some cases of male infertility are associated with disturbances of histone retention. We suggest that histones have to be in the right places in order to ensure appropriate gene expression in the early embryo and thus embryo viability. Such cases of infertility are associated with relatively gross disturbances detected as alterations of the histone:protamine ratio. It is highly likely that more subtle differences also exist and that the male population is therefore relatively heterogeneous in terms of histone retention. In a genetically diverse species such as humans, it is reasonable to suggest that the female population may be likewise varied in the tolerance of the egg to differing patterns of histones. Thus, within a single population, the pattern of histone retention represents one mechanism by which we can understand reproductive compatibility/incompatibility between individuals.

A consequence of the suggestion above is that closely related species may differ more in the epigenetic signal provided by the histones in the sperm than in the nature of their genes. Incompatibility between sperm-histone distribution and the machinery in the egg for driving gene expression would enforce reproductive isolation, even though the two forms of DNA (genotypes) themselves may be so similar as to be compatible. By this means, reproductive isolation could be ensured between variants within in a species even while they were occupying the same ecological niche and through this isolation, additional differences could be acquired and developed, driving the variants further apart until they were distinct species. It is usually assumed that reproductive isolation is the rate-limiting step in evolution, since before this occurs, new, variant phenotypes can be diluted and absorbed among the rest of the population. Isolation reduces the effective population size and evolution can then proceed more quickly. Therefore, even subtle differences in where the histones are retained could split a population into sub-groups, based on complementary differences among the females in the tolerance of the egg.

The reasons why histones may not be retained in the appropriate place on the DNA strand are not known. In the case of gross disturbances, it may well be purely a malfunction of the relevant cellular systems. In the case of more subtle defects, mutations at particular sites responsible for regulating the process could be the cause. The former scenario is more likely to be involved in the development of pathological outcomes; the latter could have a role in the establishment of reproductive isolation.

Conclusions

This brief and far from comprehensive survey of some of the ways in which the inheritance of characteristics need not be dependent on DNA reveals a couple of interesting observations. First, the relationship between pathological aberrations and evolutionary changes. It is striking that biomedical investigations looking at how the faithful inheritance of characteristics can be perturbed in the short term can yield insights into long term alterations. Thus, the molecular mechanisms leading to heritable disorders may also produce the low level of silent or, occasionally, beneficial changes that over millennia provide the diversity on which natural selection can act. The more we can understand of mechanisms leading to heritable disease, the more we may learn about the molecular motors of evolution.

Of more immediate significance is the evidence now accumulating that epigenetic factors in the germline are likely to influence gene expression in the offspring. Alterations to these factors may be the result of the direct action of environmental agents or the result of mutations in the genes controlling the epigenetic processes. Either way, it is possible that epigenesis will be found to have a much greater impact on reproductive success than mutagenesis. If the changes in the epigenetic signals are not so profound as to induce complete infertility, they may instead produce reproductive isolation within a population, or sharpen the boundaries between sub-species already starting to diverge. The splitting of the population will then increase the rate at which evolution can occur. One hundred and fifty years on from 'The Origin', Neo-Darwinism now has a new tool at its disposal, the inheritance of non-genomic variation, whose impact now needs to be assessed and integrated into our current understanding of evolution.

References

1. Trasler, J.M., Hales, B.F., Robaire, B.: Paternal cyclophosphamide treatment of rats causes fetal loss and malformations without affecting male fertility. Nature **11–17**, 144–146 (1985)
2. Brinkworth, M.H., Nieschlag, E.: Association of cyclophosphamide-induced male-mediated, foetal abnormalities with reduced paternal germ-cell apoptosis. Mutat. Res. **447**, 149–154 (2000)
3. Brinkworth, M.H.: Paternal transmission of genetic damage: findings in animals and humans. Int. J. Androl. **23**, 123–135 (2000)
4. Sorahan, T., McKinney, P.A., Mann, J.R., Lancashire, R.J., Stiller, C.A., Birch, J.M., Dodd, H.E., Cartwright, R.A.: Childhood cancer and parental use of tobacco: findings from the inter-regional epidemiological study of childhood cancer (IRESCC). Br. J. Cancer **84**, 141–146 (2001)
5. Youngson, N.A., Whitelaw, E.: Transgenerational epigenetic effects. Annu. Rev. Genomics Hum. Genet. **9**, 233–257 (2008)
6. Rassoulzadegan, M., Grandjean, V., Gounon, P., Vincent, S., Gillot, I., Cuzin, F.: RNA-mediated non-mendelian inheritance of an epigenetic change in the mouse. Nature **441**, 469–474 (2006)
7. Hammoud, S.S., Nix, D.A., Zhang, H., Purwar, J., Carrell, D.T., Cairns, B.R.: Distinctive chromatin in human sperm packages genes for embryo development. Nature **460**, 473–478 (2009)
8. Arpanahi, A., Brinkworth, M., Iles, D., Krawetz, S.A., Paradowska, A., Platts, A.E., Saida, M., Steger, K., Tedder, P., Miller, D.: Endonuclease-sensitive regions of human spermatozoal chromatin are highly enriched in promoter and CTCF binding sequences. Genome Res. **19**, 1338–1349 (2009)

Index

A
Achondroplastic dwarfism, 169
Adaptive, 3, 15, 17, 19, 33, 38, 78, 181, 191, 197–207, 213–215, 217–219, 231, 235, 243, 249
  evolution, 6, 197–206, 218, 226, 228
Affordances, 71, 202, 203
Age, 55, 73, 75–76, 88, 169, 170, 172, 178
Allele, 134, 135, 179–180, 251
Ancestorism, 149, 157–162
Angraecum sesquipedale, 93–108
Anthropic cosmological principle, 91
Anthropic epistemology, 5, 90–92
Antidepressants, 23, 24, 26, 29
Ariew, A., 132, 195, 205

B
Bacon, F., 88, 114
Baconian induction, 113–114
Bargaining, 37
Benard cells, 52
Big bang, 90–91
Biodiversity (of species), 122
Biological evolution, 34, 72, 79, 88, 92, 129, 225
Biology, 2–6, 34, 43, 48, 51, 54, 66, 69–72, 74, 80, 87, 88, 95, 122, 126, 129–144, 167, 168, 171, 173–174, 177, 178, 180, 188, 192, 197–207, 211, 225–245
Blind insects, 122–125
  spot, 50
Blindness, 14, 50, 123, 181, 235
Bohr, N., 87
Brain, 3–5, 12–21, 23, 26–28, 34, 38, 39, 43, 45–58, 60, 67–69, 71, 177, 179
  activation, 56
  imaging, 43, 55, 56
  structures, 13, 16, 50

C
Canalization, 221
Cartesian ego, 49
Cartesian self, 43, 48
Case studies, 58, 97, 108, 114, 123
Center of narrative gravity, 47
Chance, 17, 23, 25, 47, 91, 139–140, 169–171, 174, 197, 204–207, 213, 214, 221, 238, 250
Chimpanzees, 15, 35, 36, 73, 120, 154
Churchland, P.M., 50, 51, 58
Co-evolutionary, 96, 103, 108, 211
Cogito, 85
Cohen, C., 148, 149, 161, 162
Competition, 25, 36, 66, 79, 142
Competitive advantage, 132, 135, 136
Complex systems, 66, 199
Conditions of existence, 191–208
Confabulation, 50
Conscious free will, 43, 48
  self, 48–50, 54
  will, 44, 53, 54
Consciousness, 4, 11–20, 49–51, 53, 55, 56, 59, 88, 90–92
  of the self, 50, 53, 56
Consilience, 117–119, 123, 126
Constraints, 3, 18, 179, 181, 185, 219, 222, 229, 233
Contingent, 65, 151, 153, 158, 205, 206
Cooperation, 182
  games, 220, 221
Copernican revolution, 87
Creationism, 5, 90, 116–118, 122, 123, 126

M. Brinkworth and F. Weinert (eds.), *Evolution 2.0*, The Frontiers Collection,
DOI 10.1007/978-3-642-20496-8,

Credibility, 70, 122, 126
Crupain, M., 56–59
Cutaneous rabbit, 50
Cuvier, 203

D

Darwin, C., v–viii, 1–6, 16, 20, 33–34, 66, 70, 72, 73, 76, 79, 80, 92, 94–97, 100–108, 111–119, 121–127, 131–134, 139, 147, 148, 151, 163, 167–169, 171, 173, 174, 178, 181, 191–198, 205, 206, 208, 225–228, 233, 244
Darwinian, vii, 1–4, 6, 43–62, 66, 70, 71, 74, 75, 79, 80, 93, 95, 101, 102, 108, 111–127, 129, 131, 133–135, 143, 144, 149, 167, 168, 171, 174, 177, 181, 186, 191–208, 211, 212, 226, 233–235, 238, 239, 243, 244, 249–253
  account, 43–62
  explanations of evolution by natural selection, 129, 144
  models, 143
  self, 54–62
Darwinism, 1–7, 16, 18, 67, 70, 72, 73, 79, 93–95, 101, 103, 106–108, 113, 119, 127, 143, 177, 185, 191, 206, 208, 211, 225–227, 243, 244
Davies, P., 90, 91
Dawkins, R., 3, 5, 66–70, 74, 79, 80, 130, 151–155, 158, 227, 229, 238, 244
Declarative memories, 57, 58
Depression, 4, 23–29, 33–40
Descartes, 11, 85
Descent with modification, 1, 66, 72, 75, 79, 81, 191, 192, 194, 197, 206, 207
Design, 2, 5, 33, 70, 73, 90, 107, 112, 119, 122–125, 171, 179, 194, 243. *See also* Intelligent design
Determinism, 47
Development, 4, 14, 16, 18, 19, 23, 24, 29, 40, 57, 59, 66, 69, 72, 74, 88, 91, 92, 101, 130, 131, 134, 142, 144, 191, 194, 197–205, 207, 213, 215, 226–233, 235–237, 239–241, 243, 251, 252
Developmental biology, 3, 6, 171, 228–234, 236, 237, 239
  symbiosis, 237, 241
Directional evolution, 132–134
Disease, 6, 20, 23, 34, 172, 173, 175, 177–181, 183–188, 237, 240, 242, 251, 253
DNA, 67, 77, 80, 168–169, 171, 176, 227–229, 234–238, 241–244, 249–252
Dualism, 11–12, 20, 65, 233
Duhem-Quine thesis, 100, 111
Dynamic restoration, 200, 201, 205

E

Ecological developmental biology, 236, 237
Edwards, A.W.F., 158, 159
Einstein, A., 46, 94, 100
Emergent, 51, 52, 66
Emotional tone, 55
Emotions, 12, 15, 17–20, 28, 33–35, 37–39, 48, 55
Environment, 12–13, 17, 20, 21, 40, 59, 68, 73, 79, 86, 93, 101, 124, 125, 130–135, 137–139, 141–144, 174, 178, 187–188, 192, 193, 199–204, 207, 213–218, 221, 231–235, 237, 238, 240, 251
Environmental, 3, 40, 124, 134, 138, 141, 142, 180, 200, 201, 203, 215, 230, 232–235, 237–239, 242, 243, 250, 251, 253
  induction, 232
Epigenesis, 253
Epigenetic, 3, 6, 7, 16, 18, 69, 201, 236–237, 241, 250–253
Epigenetically, 16, 18
Epiphenomenalism, 18
Episodic memories, 57–59
Epistemology, 3, 5, 20, 85–108, 187, 225
Eugenics, 147, 168, 181
Evolution, v, vii, 1–7, 14, 17–19, 34, 36, 38, 39, 50, 66, 69–75, 79, 86–88, 92, 94–97, 101–103, 105, 106, 108, 113, 117, 119–122, 125, 129–134, 136, 137, 142, 144, 147, 149–151, 163, 167–178, 182, 186, 191–208, 213–216, 220, 225–235, 237, 239, 242–245, 249, 251–253
  at random, 174–176
  by natural selection, vii, 105, 106, 129–133, 136, 144, 197, 225, 226
Evolutionary, v, vii, 2–7, 15, 17, 19, 23–29, 33–40, 43, 48, 50, 51, 54, 55, 58, 59, 67, 69, 71, 72, 85–96, 103, 106, 108, 112, 119, 121, 122, 126, 129–144, 147, 177–189, 192, 194, 195, 197–199, 201, 203–207, 211–213, 218–222, 225–234, 236–243, 245, 249, 250, 252
  adaptation, 4, 23–29, 35, 39, 40
  biology, 3, 4, 48, 51, 54, 95, 180, 192, 199–201, 203, 228, 234, 240, 245
  change, 67, 71, 130, 131, 144, 192, 195, 198, 199, 201, 203, 206, 207, 243, 252
  developmental biology, 228–234

dynamics, 220–222, 236
epistemology, 3, 5, 85–92, 225
ethics, 147
explanation, 2, 33–40, 86, 87, 108, 130, 133–135, 192, 204–205
force, 135, 199
history, 180, 181
medicine, 3, 6, 177–189
models, 34–39, 88, 129, 131, 137, 144, 211
models of depression, 35–40
origins, 25, 34
process, 134–135, 140, 179, 213, 229–231, 233
stable strategies, 218, 219, 222
theory, vi, 3, 5, 6, 50, 95, 106, 108, 112, 119, 121, 122, 126, 129, 136, 137, 177, 183, 192, 195, 197, 239

Evolutionism, 75–76
Explanation, vii, 1, 2, 5, 15, 20, 26, 33–40, 52–54, 57, 66, 78, 86–88, 100, 108, 112, 114–119, 122, 125–127, 129–136, 139, 142, 144, 159, 160, 182, 185, 187, 192, 194–197, 202–205, 214, 225
Explanatory mechanism, 119, 121

F

Falsifiability, 94, 97, 98, 100–102, 105, 106, 108, 112, 113, 126
Falsificationism, 111–113, 126, 127
Family resemblances, 58, 80
Fertility, 170, 174, 251
Fine tuning, 5, 90, 91, 230
Fitness, 37, 38, 72, 93, 94, 107, 132, 134, 136, 140, 142–144, 181, 195, 215–220
Fixism, 75–76
fMRI. *See* Functional magnetic resonance imaging
Forces, 79, 80, 94–95, 99, 100, 106, 116, 121–122, 131, 134, 135, 141, 152, 195–197, 199, 200, 204, 216, 237
Free will, 4, 43–62, 71
Functional magnetic resonance imaging (fMRI), 55, 56, 59–61

G

Galapagos Islands, 174
Galton, F., 168, 174
Gazzaniga, M.S., 50, 56
Generative entrenchment, 212, 214–216
Genes, 2, 4, 16, 26, 38, 51, 66, 67, 69, 74, 77, 79, 80, 168, 171, 173–175, 179, 180, 188, 197–199, 201–204, 207, 211–222, 226, 228, 229, 232, 233, 236–238, 240–242, 250–253
Genetic damage, 169, 171, 250
drift, 169
variation, 2, 73, 175, 249
Genetic(s), 1–3, 5, 6, 38, 40, 48, 66, 69–71, 73, 79, 102, 137, 158–160, 167–169, 171, 173–175, 177, 180, 181, 200, 201, 207, 211–219, 226, 228–229, 231–245, 249, 250, 252
Genome, 6, 211–213, 227, 231, 233, 235–243, 249
Genotype, 134, 143, 198, 251, 252
Geocentrism, 119–121
Germline, 249, 250, 253
Germplasm, 198, 237–238, 242–243
Goal-directed systems, 205
Goal-directedness, 204, 205
Gödel's theorem, 88
Godfrey-Smith, P., 132, 215, 216

H

Hawkmoth, 102, 108. *See also Xanthopan morgani praedicta*
Health, 6, 172, 175, 177, 178, 181, 183–189, 237
Heatherton, T.F., 55, 56
Heredity, 4, 130, 143
Heritability, 38, 143, 249–253
Higher order effect, 192, 194–198, 206
Hodge, M.J., 195
Hoffman, D.D., 50
Homeostatic, 40, 199
Homunculus, 49
Human evolution, 6, 72, 73, 167–176, 232
Hume, 47, 85, 111
Hypothetico-deductive methodology, 5, 68, 111, 113
Hypothetico-deductive-nomological methods, 68

I

Imitation, 5, 67, 80, 139, 141
Induction, 85, 86, 111–114, 200, 226, 232, 233, 250
by elimination, 114
to the best explanation, 114
Inference, 5, 69, 103, 111–127
to the more favored, 115, 119
Infertility, 251–253

Inherency, 204–207
Inherent, 18, 91, 184, 204–207
Inheritance, 1, 3, 6, 7, 36, 40, 48, 88, 121, 143, 175, 191, 194, 198–199, 201, 202, 204, 207, 211, 232, 238, 240, 243, 249, 250, 252, 253
Innovation, 70, 71, 80, 133, 136, 139–141, 213–215, 230, 242, 243
Intelligent design, 5, 90, 112, 113, 117, 119–122, 127
Intentionality, 12, 19, 20, 62, 66, 69, 72, 194
IQ, 170

J
Johnston, V.S., 50

K
Kant, I., 75–76, 79, 87, 88
Keenan, J.P., 56–59
Keller, E.F., 199
Kelley, W.M., 55, 56
Kepler's three laws of motion, 118
Kirker, W.S., 58
Kjaer, T.W., 56–59
Klein, S., 58, 59
Knowing, 58, 78, 89, 113, 186, 233
Koch, C., 50
Kuiper, N.A., 58

L
Lakatos, I., 97–103, 106, 108, 112
Lamarck, 69, 121, 130, 131
Lamarckism, 211–212, 240
Lamarkian, 251
Lateral genetic transfer, 238–239, 242–245
Law of likelihood, 114, 115
Ledoux, J., 49, 50
Lewontin, R.C., 130, 133, 158, 159, 225
Li, T., 55, 59
Libet, B., 45, 46, 51, 53
Lisanby, S.H., 56–59
Llinas, R., 47–50, 54
Lorenz, K., 87, 88
Lou, H.C., 56–59
Luber, B., 56–59
Lyell, Sir Charles, vi–viii, 193, 202

M
Macrae, G.N., 55
Macrea, C.N., 56
Madagascar star orchid, 5, 95, 97, 103. *See also Angraecum Sesquipedale*
Malthus, T., 193
Markov process, 134
Martians, 156–157
Matthen, M., 132, 195
Mechanism(s), 4, 15, 18, 19, 25, 38–40, 49, 50, 52, 60, 66, 71, 79, 86, 107, 119, 121, 122, 126, 130, 134–144, 178, 183, 191, 194, 195, 201, 202, 204, 205, 211–214, 221, 229, 230, 232, 233, 235, 236, 240, 243, 244, 250–253
Melanin, 171, 172
Meme, 4, 5, 65–81, 130
Mendelian, 177, 198, 241
Metazoans, 198, 229, 236
Microbiome, 236, 237
Mind, 2–5, 11–21, 36–38, 43, 49, 55, 57, 69, 78, 80, 81, 87, 89–92, 112, 123, 151, 187, 195, 213, 227, 236, 240
Mind-body problem, 11, 17
Mobile genetic elements, 238, 242, 245
Mobilome, 242
Modern Synthesis Theory of evolution, 191–192, 194, 197, 199, 204, 226–228, 234–235
Monod, J., 90–92, 204
Monroe, R., 49
Moore's paradox, 88
Moral status, 149–158, 160–162
Morin, A., 55, 56
Muller, G.B., 169, 205
Multilevel selection, 183
Mutagenesis, 169, 235, 250, 253
Mutation(s), 1, 3, 4, 6, 69, 134–136, 140–141, 143, 144, 168–172, 180, 198, 200, 201, 204–207, 213–214, 218, 227–229, 231–235, 243, 249–253
  rate, 169–171, 250

N
Nagel, T., 58, 88, 89
Naïve falsifiability, 101, 105, 108
Nash equilibria(um), 218, 219
Natural selection, vi, vii, 1, 3–6, 19, 20, 34, 50, 71, 73, 88, 94, 101, 104–107, 112, 113, 115–119, 121–123, 126, 129–144, 169, 171–174, 177–179, 181, 185, 186, 188, 192–197, 204, 216, 219, 225, 226, 228, 230, 232, 238, 243, 244, 249
Natural selection explanation, 132, 136, 142–144
Neo-cortex, 88

Neo-Darwinian, 4, 6, 69, 134, 227–229, 231, 234, 235, 243, 245, 249–253
Neo-Darwinism, 2, 3, 6, 7, 177, 225–245, 249, 253
Neo-Darwinist, 66, 78, 79, 88
Neoclassical economics, 136, 137
Neptune (discovery of), 97–103, 105, 108
Neural Darwinism, 4, 16, 18
Neural nets, 58
Neurobiology of free will, 55
Neurochemical, 38–40
  theory of depression, 23
Neuroscience, 12, 38, 43–48, 54, 55, 58
Newman, S.A., 205
Newton's law/theory, 98–100, 102, 118
Niche-construction, 216, 230, 231
Niches, 123, 202, 216, 230, 231, 252
Nielson, T.I., 44, 45, 53, 54
Novelty, 130, 133–135, 139, 142–144, 230, 232, 243
  producing mechanism, 130, 135
Nowak, M., 56–59

O

Occam's razor, 26
Oderberg, D., 149–151, 155, 160–162
Organism(s), 3, 6, 48, 53, 55, 68, 74, 76, 79, 107, 140, 179, 181, 182, 185, 191, 193–195, 197–207, 211, 213–218, 228, 230–244, 250, 251
Organism-centered, 192, 201, 203, 205, 206
  approach, 206
  biology, 201–207
  environment relations, 192, 201–203
Organismal, 191, 197, 198, 201–204, 236, 237
Origin of species, v, 5, 72, 94, 96, 106, 111, 114, 121, 122, 126, 133, 191–192, 197, 205–208

P

Pain, 26, 34, 35, 39, 108, 161, 184, 187
Panpsychism, 11–12, 20
Paramutation, 250, 251
Pathology, 4, 25, 26, 250, 252
Perception of the self, 49, 56
Personality traits, 37, 38, 56–61
PET. *See* Positron-emission tomography
Phenotype, 69, 71, 143, 198, 200–202, 205, 207, 217–222, 228–232, 236, 239, 242, 251, 252
  accommodation, 200
  effects, 68, 69, 204
Phenotypic repertoire, 200
Plastic phenotypes, 221
Plasticity, 6, 17, 52, 71, 199–201, 203, 205–207, 212, 214–216, 221, 231–234, 239, 242, 244. *See also* Plastic phenotype
Platek, S.M., 55
Plato, 58, 78
Popper, K., 93–95, 97–102, 106–108, 111–114, 126, 127, 225, 230, 231, 240, 244
Positron-emission tomography (PET), 56–59
Prediction, 5, 26, 47, 94, 96–98, 100–108, 111, 112, 114, 121, 167, 221
Prior probability, 116, 117
Private language argument, 85
Probability arguments, 121, 126
Procedural memories, 57–59
Production technique, 137–141, 143
Psychology, 20, 54, 58, 68, 89, 97

Q

Qualia, 12, 17–19
Quine, W.V., 11, 13, 20, 100–101

R

Racism, 148, 150, 157–162, 168
Ramachandran, V.S., 48
Random change, 169, 174
Random mutation, 3, 81, 204–207, 213, 235, 251
Rayleigh-Benard convection, 51–52
Re-entrant activity, 16, 17, 20
Re-entry, 16–17, 19, 20
Replicator, 4, 5, 66–69, 192, 197–199, 201–207, 218, 221
  biology, 192, 197–199, 202–204, 206, 207
Reproduction, 3, 25, 26, 87–89, 92, 131, 135, 136, 143, 144, 168–171, 174, 181, 184, 185, 188, 193, 195, 197, 199, 202, 204, 206, 207
Research, 3–6, 24, 40, 59, 69, 80, 90, 95, 98, 101, 106–108, 125, 137, 139–140, 148, 149, 228–230, 232–244
Retroviral symbiosis, 241, 242
Ring species, 152–155
Robustness, 200, 226, 242
Rogers, T.B., 58
Russell, D.P., 46, 90–92, 96
Ryle, G., 58

S

Sackeim, H.A., 56–59
Saxe, R., 56, 57
SC. *See* Special creationism
Scandinavia, 172, 173
Schrodinger, E., 196
Secondary sensory qualities, 49
Selection, vii, 2–6, 15–19, 25, 34, 37–40, 51, 69, 79, 133–136, 138–144, 179, 181–185, 194, 195, 197–199, 204, 205, 212–216, 226–234, 237–239, 243–244. *See also* Natural selection
Selection for, 34, 38, 40, 122, 143, 195, 214–216
Selective advantage, 34, 37, 40
  pressures, 142, 178, 179, 235
Self, 4, 15, 19, 43–62, 85, 199
  attribution, 56, 57
  awareness, 6, 55, 148
  face recognition, 55–56
  knowledge, 43, 55, 56, 58, 59, 61
  recognition, 55
  referential, 55, 58
  relevance, 56
  relevant judgments, 56–57
Selfish genes, 66, 67
Semantic memory, 57–59
Separation distress, 4, 35, 38–40
Sexual selection, 88, 180, 181
Silberstein, M., 52
Singer, P., 148–150, 155, 161–162
Sober, E., 106, 107, 114–117, 125, 130, 131, 133, 134, 144, 195, 197, 229, 237
Social Darwinism, 72, 73
  learning, 213, 214
Social rank, 36–37
Socrates, 58
Somatoplasm, 198
Sophisticated falsifiability, 101, 105, 108
Special creationism (SC), 117, 118, 122–126
Species, vi, vii, 1–3, 6, 13, 19, 24, 26, 29, 33, 35, 38, 50, 67, 70–72, 74–77, 79, 94–96, 102, 104–106, 108, 120–124, 131–133, 135, 148–163, 171, 174, 175, 179, 184, 185, 193, 194, 197, 217, 229, 232–233, 237–240, 242–244, 252
Speciesism, 6, 148–151, 153–163
Spencer, 72, 106, 107, 130
Spinoza, 46, 91
Stag Hunt game, 221
Strong emergence, 51, 52
Struggle for life, 191–208
Subjective self, 18
Substance dualism, 11
Survival of the fittest, 72, 79, 94, 95, 101, 102, 106, 107, 163, 230
  rate, 173, 174

T

TE. *See* Thomistic epistemology
Technology, 5, 67–72, 77–81, 88, 133, 169, 220
Teleonomic, 204
Thalamus, 14, 28
Theory of evolution, 6, 86, 92, 94, 97, 103, 106, 108, 113, 147, 149, 191–192, 197, 207, 227
  forces, 131
  mind, 55, 57
  neural group selection (TNGS), 16
Third ventricle, 4, 23–29
Thomistic epistemology (TE), 89–90, 92
Time machine, 151, 167, 168
Tononi, G., 49–51
Tooley, M., 148
Transformational evolution, 130, 131
Tulving, E., 57
Turk, D.J., 56

U

Unitary self system, 56
Utopia, 167, 176

V

Values, 14, 15, 17, 18, 29, 50, 52, 60, 70–71, 117, 121, 134, 137, 138, 140, 148, 178, 184, 186–188
Variation, 2, 38, 71, 73, 76, 121, 136, 158, 169, 174, 175, 197, 201, 204, 226–231, 233, 244, 249, 250, 253
Variety, 15–17, 20, 55, 56, 73, 92, 133, 134, 138, 139, 192, 227, 228, 230, 233, 235, 237, 238, 241, 242
Vera causa, 122, 195
Vestiges, 33, 34, 36, 80, 181
Vitamin D, 172–173
Volkow, N.D., 55
von Helmholtz, H., 195

W

Wallace, A.R., 96–97, 104
Walsh, D.M., 6, 195

Walter, W.G., 44, 45, 53, 54, 97, 105
Watson, G., 47, 52, 226
Wegner, D.M., 44, 47, 53–54
Weil, S., 90
Weismann, A., 69, 198, 240
West-Eberhard, M.J., 200, 231–233, 237, 239
Whewell, W., 118
Will, 4, 43–62, 90
Williams, G.C., 177, 181

X

Xanthopan morgani praedicta, 97, 102–107. *See also* Hawkmoth
Y chromosome, 174, 175